Vom Fachbereich Mathematik
der Technischen Universtät Darmstadt
zur Erlangung des Grades eines
Doktors der Naturwissenschaften (Dr.rer.nat.)
genehmigte Dissertation

On the extraction of computational content from noneffective convergence proofs in analysis

von

Dipl.-Math. Pavol Safarik (aus Bratislava)

Referent: **Prof. Dr. Ulrich Kohlenbach**
1. Korreferent: Prof. Dr. Thomas Streicher
2. Korreferent: Prof. Dr. Benno van den Berg

Tag der Einreichung: 17. 10. 2014, Tag der mündlichen Prüfung: 17. 12. 2014
Darmstadt, D17

Bibliografische Information der Deutschen Nationalbibliothek

Die Deutsche Nationalbibliothek verzeichnet diese Publikation in der Deutschen Nationalbibliografie; detaillierte bibliografische Daten sind im Internet über http://dnb.d-nb.de abrufbar.

ISBN 978-3-8325-3765-4

Logos Verlag Berlin GmbH
Comeniushof, Gubener Str. 47,
10243 Berlin
Tel.: +49 (0)30 42 85 10 90
Fax: +49 (0)30 42 85 10 92
INTERNET: http://www.logos-verlag.de

Contents

Introduction

Foreword

In [123], a recent Fields medallist, Terence Tao, published some thoughts on the relation between "hard (quantitative, finitary)" and "soft (qualitative, infinitary)" analysis. In this essay, he also emphasized the importance of the so called "hard" analysis:

> *... I therefore feel that it is often profitable for a practitioner of one type of analysis to learn about the other, as they both offer their own strengths, weaknesses, and intuition, and knowledge of one gives more insight into the workings of the other. ...*

His point is illustrated by the proof of Theorem 1.6 in [122], which is carried out in the context of "finitary ergodic theory":

> *... The main advantage of working in a finitary setting ... is that the underlying dynamical system becomes extremely explicit. ...*

He goes on to connect the finitisation to the methods we will employ in this thesis:

> *... In proof theory, this finitisation is known as Gödel functional interpretation ...*

In the case of convergence theorems Tao calls the finitary formulation metastability and the corresponding explicit content rate(s) of metastability. It was recently observed by Avigad and Rute [10] that in the case of the von Neumann mean ergodic theorem a particular such rate of metastability (extracted in [80]) can be used to obtain even a simple effective (and also highly uniform) bound on the number of fluctuations (for the case of Hilbert spaces this was already obtained with an even better bound in [54]).
This leads us to a very natural question: What kind of effective bounds (explicit information) is extractable under which general logical conditions on convergence proofs.
In this thesis we discuss four natural kinds of such finitary information and analyze the corresponding conditions.
While the results we obtain in the process explain a very common form of realizers for strong ergodic theorems in seemingly unrelated logical circumstances, there is a notable exception to this pattern published in [107]. We will see how the nested realizer from [107] relates to a separating example for two of our kinds of finitary information mentioned above.
Once we established this basis for computational content extraction, we also investigate, so to say, the opposite approach. Namely, given the fact that one is interested in such extraction, the question arises in which systems this is still possible and how. We answer this question for a very prominent case in the last chapter, where we introduce an intrinsic formalisation of non-standard analysis and discuss the extractability of effective bounds from proofs in that system.

Scientific Context

At least since the presentation of Gödel's functional interpretation (also called Dialectica or simply "D-" interpretation) by K. Gödel in 1958, proof theoretic methods can be used not only simply to analyze the syntactic structure of proofs, but also to deliver new information about the mathematical theorems being proved. The first to formulate this idea of "unwinding proofs" was G. Kreisel:

> *What more do we know if we have proved a theorem by restricted means than if we merely know that it is true.*

Kreisel's "Unwinding of Proofs" developed into a new field of Mathematical Logic providing many new results in Algebra, Number Theory, and Numerical and Functional Analysis. To describe better the character and aim of this field, D. Scott suggested to call this discipline *Proof Mining* and the new name was quickly adopted.
In this thesis, we make mainly use of the above mentioned functional interpretation (in various forms, we will even give a new variant suitable for our formalization of non-standard analysis in the last chapter). Gödel's functional interpretation requires an *intuitionistic proof, (also called a* constructive proof), which is – roughly speaking – a proof which doesn't make any use of the law of excluded middle:

$$\mathsf{LEM} \quad : \quad \varphi \vee \neg\varphi.$$

Of course, one often works with semi-intuitionistic (semi-constructive) systems, where LEM is only partially added, e.g. only for negated formulas (i.e. formulas in the form $\neg\varphi$) – denoted by $\mathsf{LEM}_\neg$ – or for formulas with only one, existential quantifier (i.e. formulas in the form $\exists\varphi_{\mathsf{QF}}$) – denoted by Σ^0_1-LEM. An intuitionistic proof can be obtained from a classical one by applying the negative translation as a pre-processing step. The negative translation also goes back to Gödel who defined a transformation of classically provable arithmetical formulas into classically equivalent and intuitionistically provable formulas in early 30's. We will use a modified version of this interpretation as presented by Kuroda in 1951 (see definition 1.17). The application of the combination of these interpretations, i.e. first the negative translation and then the functional interpretation, is often referred to as the ND-interpretation (ND for negative D-interpretation). We will discuss all of this in great detail in the first chapter and partially in the next paragraph. For even more thorough and self-contained presentation of all these topics, see U. Kohlenbach's book [71].
In 1962, Spector [117] defined a particularly simple form of bar recursion (recursion on well-founded trees), which is sufficient to solve the functional interpretation of (negative translation of) the schema of full comprehension over numbers:

$$\mathsf{CA}^0 \ : \quad \exists f : \mathbb{N} \to \mathbb{N} \forall x \in \mathbb{N} \left(f(x) =_{\mathbb{N}} 0 \leftrightarrow \varphi(x) \right).$$

One could call this result a major break-through for proof mining, since it suffices to prove most of classical analysis within a simple logical system based on Peano arithmetic (again see also [71]).
In particular, with Spector's bar recursion one can extract realizers (using ND-interpretation) for almost all theorems of classical analysis (given their formal proof).
In fact, for many theorems the existence of a uniform bound on the realizers is guaranteed by Kohlenbach's metatheorems introduced in [70] and refined in [37]. Additionally, proof theoretic methods such as Kohlenbach's monotone functional interpretation (see [62]) can be used to systematically obtain such effective bounds from their proofs, which tend to be a lot simpler than the actual realizers, though they remain equally useful (again [71] is a very good reference for a comprehensive discussion on this topic).
We use Spector's solution together with Kohlenbach's monotone interpretation of Weak König's Lemma (WKL, see 1.61 in first chapter or for more details and background e.g.

[61, 71, 53, 126]) to give the ND-interpretation of the Bolzano-Weierstrass ($\mathsf{BW}_{\mathbb{R}}$) theorem (see Definition 4.1, where we give a proper formalization for a more general space):

Any bounded sequence in $\mathbb{R}^d$ has a limit point.

The computational contribution of (instances of) BW and other principles describing sequential compactness has been analyzed by Kohlenbach in [64] for proofs in somewhat weaker base systems: the Grzegorczyk arithmetic of level n, $\mathsf{G}_n\mathsf{A}^\omega$. Kohlenbach and Oliva gave bounds for the bar-recursive realizers of the Principle of monotone convergence, PCM, in the last section of [82] and discuss various general scenarios, where PCM has provably significantly smaller computational contribution.
Still, it is a bit surprising that the computational contribution of this basic theorem was first fully (in the above sense) analyzed in [106]. Thereafter the interest in BW's computational content continued. It was analyzed as part of two different larger projects. Encouragingly, as far as the results can be compared, in both cases the computational content corresponds very closely to the one obtained in [106].
The more recent of these results were obtained by P. Oliva and T. Powell in [100], where the authors take the same approach as in [106]. The main difference is, that Spector's bar recursion is replaced by the use of selection functions. The authors show the equivalence of unbounded products of selection functions and bar-recursion in [101] and demonstrate how this can be used to give a different formalization of the realizers in [106]. Moreover, they give a very nice game-theoretic interpretation of the realizers which allows for better understanding and intuition for he original bar-recursive solutions.
On the other hand, V. Brattka analyzed the BW theorem as part of the Weihrauch-Lattice (see e.g. [16, 47]) in [17]. Though using completely different techniques, also he arrives at the conclusion that while using BW in its full generality in a proof implies very strong forms of recursion, when used in a more natural way, i.e. applied as a single instance of the principle, the complexity is rather low.
Finally, we should mention that in a formulation of BW where one claims only the existence of a Cauchy subsequence (as opposed to a fast converging subsequence), this principle becomes equivalent to the much weaker cohesive principle (see [88]).
In general, the proof mining experience shows that very often proofs of well known and influential theorems do not need this kind of strong principles, i.e. principles like CA^0 requiring more than normal recursion (though in most cases the proofs themselves are formulated in ways, which would suggest otherwise).
In [8], J. Avigad, P. Gerhardy and H. Towsner analyzed the mean ergodic theorem, implicitly showing that, when proving that the sequence of averages is Cauchy (rather than its full convergence), it can be proved with strictly arithmetical principles[1], i.e. statements about numbers without the need of a specific true higher type object like an actual cluster point for a sequence as in the BW theorem (please refer to section 3.1 for more details on arithmetization in this context and to [64] in general).
Shortly after [8], Kohlenbach and Leuştean gave a more efficient analysis of MET in [80], which was further investigated by Avigad and Rute in [10], where they derive the above discussed bound on the number of fluctuations (roughly speaking a bound on how many times does the distance between the values of the sequence exceed a given ϵ, see Definition 2.2). Such a bound gives us, of course, much more than simply a rate of metastability. The question under which circumstances can a fluctuation bound be (systematically) obtained – hoping for a similarly clear-cut answer as in the case of

[1] more precisely, with arithmetic principles and arithmetic versions of analytical and higher principles

metastability – was answered in [83] and led to a new but very natural and straightforward definition of (effective) learnability. As expected, the kind of computational information we get, strongly depends on the "amount of classicallity" we use. Or in other words, how much of LEM was truly needed to prove a given theorem.
Interestingly, to some extent, we get a bit of "classicallity" in constructive formalization(s) of non-standard analysis as well (see e.g. [9]). Of course, this makes logical formalization of non-standard analysis and possible related conservation results interesting on its own account. Moreover, it would be hugely beneficial, to have an extraction procedure and some meta-theorems to guarantee specific uniform bounds in the non-standard context as well. Inspired by the work of Nelson [97, 98], Berg et al. give a formalization of non-standard analysis together with an adapted form of Shoenfield interpretation (while the similarity between Shienfield's interpretation and Nelson's translation procedure between classical and non-standard proofs was actually the original point of departure), which are both intended for extraction of computation content from proofs based on non-standard methods, in [14]. A major step towards that goal would be solving the interpretation of the saturation principles, which can be used to formalize the concept of Loeb measures – one of the most prominent non-standard techniques. For certain systems, it has turned out that extending them with saturation principles has resulted in an increase in proof-theoretic strength (see [45, 56]). So far, it seems that in the context of [14] it is also the case and we need to add a version of bar recursion which corresponds precisely to Spector's, but adopted to our special form of sequence application as defined in [14].

Goals and Organization of the Material

While in the context description above, we follow more or less the chronological order of development, this thesis is structured according to the underlying proof-theoretic strength or, correspondingly, the complexity of the computational content in question. We start with low complexity realizers for simple convergence and work towards bar-recursive solutions in the last two chapters. Large parts of this thesis are based on previous publications. We precede each section with a short quote of some of the journal's referee comments on the corresponding article. These quotes are intended to offer a short abstract on the topic at hand from a more detached perspective. In the following we give an overview of the contents of the chapters, their relation and which publications are they based on.
The *Preliminaries chapter* is somewhat different from the others. It is based on parts from both a joint paper with U. Kohlenbach [108] and [106]. It gives a quick introduction to the basic methods and results in the area of proof mining, which might be considered folklore or common knowledge. We do give some new results (and fill tiny gaps in known proofs), but mostly it is a collection of known or not surprising material and can be found in standard literature (mainly [125, 53, 71]). In fact, it is meant more as a quick reference for results we need in later chapters. We capture the most relevant results in short summaries at the end of the corresponding sections. For an interested reader, it is probably best to refer to [71] directly instead.
The *second chapter* is about effective learnability and discusses in great detail what kind of computational content can be extracted, and under which circumstances, from convergence proofs. It is based on the most recent work with U. Kohlenbach [83] to which both authors contributed equally [2] and which was written up by myself (except for

[2]Though the proof of Proposition 2.33 is solely due to Kohlenbach.

the introduction to [83] and some smaller remarks not reproduced here). We start with this topic, because the analysis deals with the lowest level of computational complexity (compared to the remainder of this thesis). The goal is to precisely characterize when can we obtain more than a (computable) rate of metastability from a proof of a convergence statement in analysis. Of course, it is known when (computable) full rate of convergence is extractable. Here, we are concerned with computational content strictly in between. Namely a (computable) bound on fluctuations or at least an (effective) learning procedure for the limit point.
We take this, so to say, a step further in the *third chapter*, where we dive into a case study on a proof of a strong non-linear ergodic theorem due to R. Wittmann. The situation here, closely resembles our abstract separating example for a convergence statement which is non-learnable but has an effective rate of metastability from second chapter. This is due to a specific nested structure of the extracted realizer, which is unprecedented by similar results in proof mining so far. It is a rare example, where a proof of strong convergence doesn't lead to a realizer of the specific form which is typically guaranteed for effectively learnable statements. This chapter is based on [107] extended by an in depth analysis of an arithmetized version of Wittmann's proof.
In the *fourth chapter* we go even further and investigate how a specific compactness principle, the theorem of Bolzano-Weierstrass, contributes to the complexity of realizers when used as the main principle in proofs of logical statements in a particular, but rather general, logical form. To achieve this, we need to use advanced techniques like bar-recursion, but we demonstrate that even in systems formalizing large parts of full classical analysis, powerful principles like BW contribute by the minimal amount possible, when they are used naturally and are treated correctly. The results in this chapter improve, generalize and extend preliminary results going back to my Diplom-Thesis [106] and have been published in [108]. In particular, we now treat the Bolzano-Weierstrass principle for general Polish spaces of the form $\mathbb{P}^b$ (see Definition 4.2) which subsequently has been crucially used by U. Kohlenbach in his proof-theoretic analysis of weak compactness in [75].
The *last chapter* surveys my joint work with B. van den Berg and E. Briseid in [14] presenting in detail my contributions (sections 5.4 and 5.5), here actually extended by an example (interpretation of DNSst). While in the chapters three and four we use established techniques already fine-tuned for proof-mining, here we want to contribute to the field in general and provide extraction procedures for non-standard analysis. Such a challenging task certainly needs more than a single publication on a system formalizing the non-standard methods and a corresponding (negative) functional interpretation, but we believe we made a very solid first step. In this thesis we go actually further and discuss a yet unpublished successful part of a larger unfinished attempt on interpreting the principle of countable saturation, which is the basic step needed for interpreting proofs based on Loeb measures (see section 5.6). This part is of particular interest in the context of this thesis, since it was achieved using much of the ideas and techniques needed for our previous results.

Notation and common Expressions

By "$\equiv$" we mean the syntactical identity. We will write Π^0_1 and Σ^0_1 for the purely universal arithmetic formulas, i.e. $\forall n \in \mathbb{N}\ \varphi_{\mathsf{QF}}(n)$, and the purely existential arithmetic formulas, i.e. $\exists n \in \mathbb{N}\ \varphi_{\mathsf{QF}}(n)$, where in both cases φ_{QF} denotes a quantifier-free formula, which may contain parameters of arbitrary type. However, in general, we allow quantification over variables of any finite type (see Definition 1.1).

For the encoding of a given finite sequence s of natural numbers we write $\mathrm{lh}(s)$ for the length of s and denote by $[s]$ the type one function:

$$[s](i^0) :=_0 \begin{cases} s(i) & \text{if } i <_0 \mathrm{lh}(s) \\ 0 & \text{else} \end{cases} .$$

For a type one function f and a natural number n we define the corresponding encoding of the finite sequence $\overline{f}n$ of length n as follows:

$$\overline{f}n := \langle f(0), f(1), \cdots, f(n-1) \rangle.$$

Given two finite sequences s and t we write $s * t$ for the concatenation of s and t. We write shortly $s * \langle 0 \rangle$ and $s * \langle 1 \rangle$ as $s * 0$ and $s * 1$. Following the notation of Avigad and Feferman [7] by $s \subseteq t$ we mean t is an extension of s (i.e. the sequence t starts with the sequence s, or is s). We denote the empty sequence by $\varnothing$.
For finite tuples of variables (not necessarily of the same type) $x_1, x_2, \ldots, x_k$ we write $\underline{x}$. By $\underline{x}^{\underline{\rho}}$ we mean $x_1^{\rho_1}, x_2^{\rho_2}, \ldots, x_k^{\rho_k}$.
Generally we will use the Greek letters φ, ψ, χ to denote formulas (as an exception a predicate in section 4 is named I), the lower case Latin letters f,g,h for functions, the letters a,b,i,j,k,... for natural numbers and encodings, and the capitals like A,B,... for functionals.
We denote $\lambda n^0.1^0$, $\lambda f^1.1^0$ and so on by $\mathbf{1} \equiv \mathbf{1}^1$, $\mathbf{1}^2$ and so on. We use bold numbers to indicate the type level of a term, e.g. we would write $t^{\mathbf{1}}$ for $t^{1(0)}$. In general, we use this superscript as a shortcut for a specific type having the given type level. So, by $\forall X^{\mathbf{2}}$ we mean for all X of an appropriate type, e.g. $2(1(0))$, not for all X which are of any type with level 2.
We will write "holds classically/intuitionistically" or "is equivalent classically/intuitionistically" and so on meaning in fact "is provable" or "the equivalence is provable" and so on in the classical, i.e. in $\mathsf{WE\text{-}PA}^\omega + \mathsf{QF\text{-}AC}$, or intuitionistic, i.e. in $\mathsf{WE\text{-}HA}^\omega + \mathsf{M}^\omega + \mathsf{AC}$, system respectively.
We will treat tuples and pairs somewhat special in the second chapter, but we will clarify this in its first section.
Finally, in our last chapter on non-standard systems, we will need some more elaborate conventions, however we define those there at the beginning, since we don't need them anywhere else.

Acknowledgments

First of all, I would like to thank my supervisor Prof. Dr. Ulrich Kohlenbach. Professor Kohlenbach not only proposed the topic for this thesis and supported my research in a very flexible way depending on the most recent results, but also went out of his way and accorded me his time whenever I encountered a problem providing hints, explanations and encouragement. Moreover, he led me to a large field of applied logic far out of the bounds of the mere subject at hand.
I am also grateful to his assistants Dr. Eyvind Briseid, Dr. Alexander Kreuzer and Dr. Laurentiu Leustean for their direct advice as well as for pointing me to some additional literature. I thank Prof. Dr. Benno van den Berg and Prof. Dr. Thomas Streicher for several fruitful discussions and tips.
I also appreciate the continuing support of my family and friends.

Preliminaries

... The paper is technically sound, and very well written. It also gives a nice overview of all the related issues around not only BW, but also WKL, majorisability, dialectica interpretation, comprehension and bar recursion ...

– anonymous referee about [108], Nov 23, 2009.

Systems

Before we can proceed with a brief introduction to proof interpretations, we need to specify the systems in which we formalize our proofs. We will restrict ourselves mostly to systems with so called weak extensionality, as the full extensionality rule can not be interpreted by functional interpretation. We will use mainly the following two base systems: the (weakly) extensional Heyting arithmetic in all finite types, (W)E-HA$^\omega$ - see definition 1.3, as the intuitionistic base system and (weakly) extensional Peano arithmetic in all finite types, (W)E-PA$^\omega$ - see definition 1.8, as the classical base system.
As their names suggest, both systems use all finite types:

Definition 1.1. The *set of all finite types*, $\boldsymbol{T}$, is generated inductively by

$$\begin{aligned} &0 \in \boldsymbol{T}, \\ \rho, \tau \in \boldsymbol{T} \to \quad &\tau(\rho) \in \boldsymbol{T}, \end{aligned}$$

where the type 0 is the type corresponding to the natural numbers $\mathbb{N}$. Objects of type $\tau(\rho)$ are functions mapping objects of type ρ to objects of type τ. We will sometimes omit brackets when specifying types, i.e. write $\rho(\tau\sigma)$ meaning in fact $\rho(\tau(\sigma))$.
The *set of pure types*, $\boldsymbol{P}$, is generated inductively by

$$\begin{aligned} &0 \in \boldsymbol{P}, \\ \rho \in \boldsymbol{P} \to \quad &0(\rho) \in \boldsymbol{P}. \end{aligned}$$

We often denote pure types by natural numbers, with 1 for 00, 2 for $0(00)$, and $n+1$ for $0(n)$.
The *degree* or *type level* of ρ is defined by

$$\begin{aligned} deg(0) &:= 0, \\ deg(\tau(\rho)) &:= \max(deg(\tau), deg(\rho)+1). \end{aligned}$$

We will use numbers in bold font for the type level of terms. I.e. for a functional A of type $(0 \to 0) \to ((0 \to 0) \to ((0 \to 0) \to 0))$ we write also $A^{\mathbf{2}}$, expressing that A has degree 2, instead of $A^{(21)1}$, where the type of A is declared precisely.
We define higher order equality as follows:

Definition 1.2. Two terms of type ρ are equal, iff:

$$s^\rho =_\rho t^\rho :\equiv \forall y_1^{\rho_1}, \dots, y_k^{\rho_k}(sy_1 \dots y_k =_0 ty_1 \dots y_k),$$

where $\rho = 0(\rho_k)\dots(\rho_1)$.

In addition, we always allow the use of purely universal axioms as well as the *quantifier-free axiom of choice*,

$$\mathsf{QF\text{-}AC} : \quad \forall x^\rho \exists y^\tau \varphi_{\mathsf{QF}}(x, y) \ \to \ \exists g^{\tau(\rho)} \forall x^\rho \varphi_{\mathsf{QF}}(x, gx),$$

in the classical case, and we always allow the *Markov principle* in all finite types,

$$\mathsf{M}^\omega : \quad \neg\neg\exists x^\rho \varphi_{\mathsf{QF}}(x) \ \to \ \exists x^\rho \varphi_{\mathsf{QF}}(x),$$

and full AC in the (semi-) intuitionistic case. In both schemas, QF-AC and M^ω, the formula φ_{QF} is quantifier-free and ρ and τ are arbitrary finite types (see definition 1.1). We obtain full AC from QF-AC by dropping the restriction on the formula φ_{QF} to be quantifier-free.
For such a system SYS we will introduce also the weakening $\widehat{\mathsf{SYS}}\restriction$, where we allow only type-0-recursion and quantifier-free induction. In this context we often add a "precise amount of induction" by adding the schema:

$$\Sigma_n^0\text{-}\mathsf{IA} \quad : \quad \varphi(0) \wedge \forall x^0(\varphi(x) \to \varphi(x+1)) \ \to \ \forall x^0 \varphi(x),$$

where $\varphi \in \Sigma_n^0$, i.e. φ is of the form $\underbrace{\exists x_1^0 \forall x_2^0 \dots}_{n} \varphi_{\mathsf{QF}}(x_1, x_2, \dots, x_n^0)$ for a quantifier-free formula φ_{QF}. To allow arbitrary many quantifiers over type 0, we write Σ_∞^0.
We denote the quantifier-free term calculus of $\mathsf{WE\text{-}HA}^\omega$ by $\mathcal{T}$. The fragments $\mathcal{T}_n$ correspond to $\mathcal{T}$ with the recursor R_ρ restricted to the types ρ of level $\leq n$. See definitions 1.83 and 1.85.

Proof Interpretation

Generally, using a suitable proof interpretation, not necessarily the D-interpretation, in the context of proof mining we can expect to extract programs and/or computable bounds from a given proof. However, the two interpretations we will use throughout this thesis, namely negative translation and D-interpretation, were originally presented with a different aim. The proof-theoretic gain of negative translation as a tool for a translation from classical to intuitionistic systems is clear. The functional interpretation was first introduced by Gödel to obtain consistency results relative to $\mathcal{T}$ motivated by Hilbert's program and not as a tool for proof mining. We give a very brief description of these interpretations right now, see e.g. [71] or [7] and definitions 1.17 and 1.26 for further details.
As mentioned earlier, we will make use of Kuroda's negative translation, which assigns to every formula φ a logically equivalent formula $\varphi' :\equiv \neg\neg(\varphi^*)$, where φ^* is just φ with

a double negation behind each $\forall$-quantifier.
Gödel's functional interpretation assigns to every formula φ an intuitionistically equivalent formula $\varphi^D :\equiv \exists \underline{x} \forall \underline{y} \varphi_D(\underline{x}, \underline{y}, \underline{a})$, where $\underline{x}$ and $\underline{y}$ are tuples of variables of arbitrary finite type, $\underline{a}$ is a tuple of parameters of arbitrary finite type and $\varphi_D(\underline{a})$ is quantifier-free. Moreover, by the soundness of the functional interpretation we know there is a tuple $\underline{t}$ of closed terms of our intuitionistic system, such that $\forall \underline{y} \varphi_D(\underline{t}\,\underline{a}, \underline{y}, \underline{a})$ holds intuitionistically and even provably in a quantifier-free calculus, provided that φ is provable in WE-HA$^\omega$ + M$^\omega$ + AC.
We should mention that under the combination of both interpretations (i.e. under the ND-interpretation) for every formula provable in WE-PA$^\omega$ with the help of quantifier-free axiom of choice and any additional universal axioms (usually denoted as WE-PA$^\omega$ + QF-AC + $\mathcal{U}$) the interpreted formula can be proved in WE-HA$^\omega$ + $\mathcal{U}$ using the same purely universal axioms $\mathcal{U}$ as we allowed in the original classical proof. Kohlenbach introduced in [62] the *monotone functional interpretation* or shortly *MD-interpretation* (see 1.8). This interpretation is based on the D-interpretation and Howard's majorizability property (see definition 1.27). Instead of delivering a concrete realizer as before, by the MD-interpretation we can construct a tuple of closed terms $\underline{t}$ which *majorize* some realizers. Unless combined with negative translation, we still need an intuitionistic proof, but, fortunately, it can use some other ineffective principles which have a direct MD-interpretation (e.g. Weak König's lemma, for general definition of these principles see theorem 1.29).

The first two interpretations in this section as well as the particular axiomatization of intuitionistic logic and arithmetic go originally back to Gödel (see [38], [39]), the monotone functional interpretation was first introduced by Kohlenbach (see [62]). The calculus of WE-HA$^\omega$ and Heyting Arithmetic in general was further analyzed by Troelstra in [125].
The herein presented description of the formal systems and representations is based mainly on [71] and we use also the shortened version extracted by E. M. Briseid (see [19]). We define the system WE-HA$^\omega$ directly with its language, axioms, and rules rather than going the usual way of defining an intuitionistic calculus for first order logic followed by the addition of arithmetic and types.

1.1 The System WE-HA$^\omega$

We call the system WE-HA$^\omega$, which we describe below, weakly extensional intuitionistic ('Heyting'-) arithmetic in all finite types.

Definition 1.3. *Weakly extensional arithmetic in all finite types* WE-HA$^\omega$

- *The language* $\mathcal{L}$(WE-HA$^\omega$) *of* WE-HA$^\omega$:
  - *The signature of* $\mathcal{L}$(WE-HA$^\omega$):
    * We use the following logical symbols (for any type ρ):

$\wedge$	and	$\vee$	or
$\to$	implies	$\perp$	absurdity
$\exists\, x^\rho$	there exists	$\forall\, x^\rho$	for all.

    * We write (for any type ρ) $x^\rho, y^\rho, z^\rho, ...$ for variables of type ρ (bound or free).

* We include the constants:

0^0	natural number 0	S^{00}	successor
$\Sigma_{\delta,\rho,\tau}$	combinator of type $\tau\delta(\rho\delta)(\tau\rho\delta)$	$\Pi^{\rho\tau\rho}_{\rho,\tau}$	projector
$\underline{R}_{\underline{\rho}}$	simultaneous recursors of type $\underline{\rho}(\underline{\rho}0\underline{\rho})\underline{\rho}0$.		

* and the binary predicate constant $=_0$ for equality between type 0 objects.

- *Terms of* $\mathcal{L}(\text{WE-HA}^\omega)$:
  1. For any type ρ, the terms of type ρ are the variables $x^\rho, y^\rho, \ldots$ and the constants $c_1^\rho, c_2^\rho, \ldots$
  2. If $t^{\tau\rho}$ is a term of type $\tau\rho$ and s^ρ is a term of type ρ, then $t^{\tau\rho}(s^\rho)$ is a term of type τ.
- *(Prime) Formulas of* $\mathcal{L}(\text{WE-HA}^\omega)$:
  1. Absurdity, $\perp$, is a (prime) formula.
  2. If s^0 and t^0 are terms of type 0 then $s =_0 t$ is a (prime) formula. Prime formulas are also called *atomic formulas*.
  3. If φ and ψ are formulas and x^ρ is a variable of type ρ then

$$\varphi \wedge \psi \qquad \varphi \vee \psi$$
$$\varphi \to \psi \qquad \exists x^\rho\ \varphi(x)$$
$$\forall x^\rho\ \varphi(x)$$

are formulas.

- *Abbreviations*:

$$\neg\varphi :\equiv\ \varphi \to \perp \qquad \varphi \leftrightarrow \psi :\equiv\ (\varphi \to \psi) \wedge (\psi \to \varphi).$$

- *Axioms of* WE-HA$^\omega$
  1. Axioms of contraction:

$$\varphi \vee \varphi \to \varphi \qquad \varphi \to \varphi \wedge \varphi.$$

  2. Axioms of weakening:

$$\varphi \to \varphi \vee \psi \qquad \varphi \wedge \psi \to \varphi.$$

  3. Axioms of permutation:

$$\psi \vee \varphi \to \varphi \vee \psi \qquad \varphi \wedge \psi \to \psi \wedge \varphi.$$

  4. Ex falso quodlibet axiom:

$$\perp \to \varphi.$$

  5. Quantifier axioms:

$$\forall x \varphi(x) \to \varphi(t) \qquad \varphi(t) \to \exists x \varphi(x),$$

  where t is free for x in φ.

6. Axioms of equality for $=_0$ (x^0, z^0 and y^0 are variables of type 0):

$$x =_0 x \qquad x =_0 y \to y =_0 x$$
$$x =_0 y \wedge y =_0 z \to x =_0 z.$$

The reflexivity, symmetry and transitivity of higher type equalities $=_\rho$ are derivable from these axioms (see lemma 1.2).

7. Successor axioms:

$$S(x^0) \neq_0 0$$
$$S(x^0) =_0 S(y^0) \to x =_0 y.$$

We write also $x+1$ for $S(x)$.

8. Induction schema for an arbitrary formula $\varphi(x^0)$ of WE-HA$^\omega$:

$$\mathsf{IA} \ : \ \varphi(0) \wedge \forall x^0(\varphi(x) \to \varphi(x+1)) \ \to \ \forall x^0 \varphi(x).$$

9. Axioms for $\Pi_{\rho,\tau}$, $\Sigma_{\delta,\rho,\tau}$ and $\underline{R}_{\underline{\rho}}$:

$$\begin{aligned} (\Pi) &: \ \Pi_{\rho,\tau} x^\rho y^\tau =_\rho x^\rho \\ (\Sigma) &: \ \Sigma_{\delta,\rho,\tau} x^{\tau\rho\delta} y^{\rho\delta} z^\delta =_\tau xz(yz) \\ (\underline{R}_{\underline{\rho}}) &: \ \text{for } i = 1, ..., k: \\ & \begin{cases} (R_i)_{\underline{\rho}} 0 \underline{y}\,\underline{z} =_{\rho_i} y \\ (R_i)_{\underline{\rho}} (Sx^0) \underline{y}\,\underline{z} =_{\rho_i} z_i (\underline{R}_{\underline{\rho}} x \underline{y}\,\underline{z}) x \end{cases}, \end{aligned}$$

where $\underline{\rho} = \rho_1, ..., \rho_k$, $\underline{\rho}^t := \rho_k \dots \rho_1$, $\underline{y} = y_1, ..., y_k$ and $\underline{z} = z_1, ..., z_k$ with y_i of type ρ_i and z_i of type $\rho_i 0 \underline{\rho}^t$.

- *Rules of* WE-HA$^\omega$

1. Modus ponens and syllogism:

$$\frac{\varphi \quad \varphi \to \psi}{\psi} \qquad \frac{\varphi \to \psi \quad \psi \to \chi}{\varphi \to \chi}.$$

2. Exportation and importation:

$$\frac{\varphi \wedge \psi \to \chi}{\varphi \to (\psi \to \chi)} \qquad \frac{\varphi \to (\psi \to \chi)}{\varphi \wedge \psi \to \chi}.$$

3. Expansion:

$$\frac{\varphi \to \psi}{\chi \vee \varphi \to \chi \vee \psi}.$$

4. Quantifier rules (assuming x is not free in ψ):

$$\frac{\psi \to \varphi(x)}{\psi \to \forall x\, \varphi(x)} \qquad \frac{\varphi(x) \to \psi}{\exists x\, \varphi(x) \to \psi}.$$

5. Quantifier-free rule of extensionality:

$$\text{QF-ER} \quad : \quad \frac{\varphi_{\text{QF}} \to s =_\rho t}{\varphi_{\text{QF}} \to r[s] =_\tau r[t]},$$

where s^ρ, t^ρ and $r[x^\rho]^\tau$ are terms of WE-HA$^\omega$, φ_{QF} is a quantifier-free formula and $\rho, \tau \in \boldsymbol{T}$ are arbitrary.

Remark 1.4. A careful reader observed that the congruence axioms for $=_0$ are not explicitly included in the axioms of equality (6). However, it is implied by QF-ER that:

$$\frac{x =_0 y \;\to\; x =_0 y}{x =_0 y \;\to\; r[x] =_\rho r[y]} \text{ QF-ER}.$$

The system obtained by the axioms 1-5 together with the rules 1-4 restricted to closed formulas corresponds to the one suggested by Gödel for the purpose of verifying the soundness of the D-interpretation. Troelstra shows in [125] the equivalence between Gödel's and Spector's system as defined originally by Spector in [117] which, in turn, was proved to be equivalent to Kleene's formalization given in [59] by Spector in [117].
Moreover, Troelstra shows the equivalence of Spector's system to the well known natural deduction calculus proving the deduction theorem for Spector's system. Both, the proof and the definition, are given in [125]. In that book, Troelstra also derives some additional schemata and rules, e.g. the contraposition rule, $(\varphi \to \psi) \to (\neg\psi \to \neg\varphi)$, or the equivalence of single/triple negation, $\neg\neg\neg\varphi \leftrightarrow \neg\varphi$.
This means that as WE-HA$^\omega$ is an extension of Gödel's system it surely proves all that can be proved by the natural deduction system as defined in [125]. As a facile motivation, we will derive a simple lemma in the form we will need later. To do so, we use Troelstra's results and the axioms and rules above. Generally, one can prove even

$$\neg\neg\exists x\, \varphi(x) \;\leftrightarrow\; \neg\forall x\, \neg\varphi(x),$$

again, using only Spector's system (or, equivalently, the selection of axioms and rules of WE-HA$^\omega$ as given above).

Lemma 1.5.

$$\text{WE-HA}^\omega \quad \vdash \quad \neg\neg\exists x\, \varphi(x) \;\to\; \neg\forall x\, \neg\varphi(x).$$

Proof. First, we derive $\exists x\varphi(x), \forall x\neg\varphi(x) \vdash \bot$:

$$\dfrac{\exists x\varphi(x)^{\text{ass.}} \qquad \dfrac{\dfrac{\forall x\neg\varphi(x)^{\text{ass.}} \quad \forall x\neg\varphi(x) \to \neg\varphi(x)^{\text{Ax. 5}}}{\varphi(x) \to \bot}\text{ Rule 1}}{\exists x\varphi(x) \to \bot}\text{ Rule 4}}{\bot}\text{ Rule 1},$$

which implies $\exists x\varphi(x) \to \neg\forall x\neg\varphi(x)$ by the deduction theorem proved by Troelstra in [125]. By contraposing this implication twice, again using [125], we obtain:

$$\neg\neg\exists x\varphi(x) \to \neg\neg\neg\forall x\neg\varphi(x),$$

what is equivalent to

$$\neg\neg\exists x\varphi(x) \to \neg\forall x\neg\varphi(x),$$

by [125] yet again. □

We have that WE-HA$^\omega$ allows λ-abstraction in the following sense.

Lemma 1.6. (see e.g. [71]) *If $t[x^\rho]^\tau$ is a term of* WE-HA$^\omega$*, one can construct in* WE-HA$^\omega$ *a term $\lambda x^\rho.t[x]$ of type $\tau\rho$ such that*

$$\text{WE-HA}^\omega \vdash (\lambda x^\rho.t[x])(s^\rho) =_\tau t[s],$$

and such that the free variables of $\lambda x^\rho.t[x]$ are the same as the free variables of $t[x^\rho]^\tau$ only without x^ρ.

We also note that prime formulas are decidable in WE-HA$^\omega$. In addition, we will be able to treat any quantifier-free formula $\varphi_{\text{QF}}(\underline{x})$ containing only $\underline{x}$ free as a prime formula, since we can find closed terms t of WE-HA$^\omega$ such that

$$\text{WE-HA}^\omega \vdash \varphi_0(\underline{x}) \leftrightarrow t\underline{x} =_0 0.$$

We can also code a finite tuple of variables of different types into a single variable, whose type depends on the types of the variables in the tuple. This can be shown in WE-HA$^\omega$ (see [125]). So, in principle, simultaneous recursion $\underline{R}_{\underline{\rho}}$ can be reduced to R_τ for suitable type τ.
The inequality $\leq$ between numbers is definable in WE-HA$^\omega$ as a quantifier-free predicate.

Definition 1.7. The inequality predicate $\leq_\rho$ between functionals of type ρ is defined inductively by

$$x \leq_0 y :\equiv x \leq y \tag{1}$$
$$x \leq_{\tau\rho} y :\equiv \forall z^\rho (xz \leq_\tau yz). \tag{2}$$

1.2 Fragments and Extensions of WE-HA$^\omega$

The classical system corresponding to WE-HA$^\omega$ is obtained simply by adding the law of excluded middle:

Definition 1.8. By adding the law of excluded middle

$$\text{LEM} \quad : \quad \varphi \vee \neg\varphi,$$

to the axioms of WE-HA$^\omega$ for any formula in WE-HA$^\omega$ we obtain WE-PA$^\omega$.

Furthermore, we define the extensional systems:

Definition 1.9. By replacing the rule QF-ER by the *axioms of higher type extensionality*

$$\text{E}_\rho \quad : \quad \forall z^\rho, x_1^{\rho_1}, y_1^{\rho_1}, \ldots, x_k^{\rho_k}, y_k^{\rho_k} \left(\bigwedge_{i=1}^{k} (x_i =_{\rho_i} y_i) \to z\underline{x} =_0 z\underline{y} \right),$$

in the systems WE-HA$^\omega$/WE-PA$^\omega$, where $\rho = 0\rho_k \ldots \rho_1$, we obtain the systems E-HA$^\omega$/E-PA$^\omega$.

On the other hand, sometimes it is sufficient to have weaker systems. This leads us to the definition of the following fragments:

Definition 1.10. For all four systems WE-HA$^\omega$, WE-PA$^\omega$, E-HA$^\omega$ and E-PA$^\omega$ let correspondingly $\widehat{\text{WE-HA}}^\omega\!\restriction$, $\widehat{\text{WE-PA}}^\omega\!\restriction$, $\widehat{\text{E-HA}}^\omega\!\restriction$, $\widehat{\text{E-PA}}^\omega\!\restriction$ denote the fragments where we only have the recursor R_0 for type-0-recursion and the induction schema is restricted to the schema of quantifier-free induction:

$$\text{QF-IA} \quad : \quad \Big(\varphi_{\text{QF}}(0) \wedge \forall n^0 (\varphi_{\text{QF}}(n) \to \varphi_{\text{QF}}(n+1))\Big) \to \forall n^0 \varphi_{\text{QF}}(n),$$

where φ_{QF} is a quantifier-free formula and may contain parameters of arbitrary types.

1.3 The ordered Field of rational Numbers within WE-HA$^\omega$

Rational numbers are represented by codes $j(m,n)$ of pairs (m,n) of natural numbers. Typically j would be just the Cantor pairing function, but any common surjective pairing function j with the inverse functions $j_1(j(m,n)) = m$ and $j_2(j(m,n)) = n$ does the job. We need the surjectivity to be able to conceive each natural number as a code of a uniquely determined rational number. For more examples on paring functions see e.g. [99].

Definition 1.11. *Coding of rational numbers*
The rational number represented by $j(m,n)$ is

$$j(m,n) \sim \begin{cases} \frac{\frac{n}{2}}{m+1} & \text{if } n \text{ is even} \\ -\frac{\frac{n+1}{2}}{m+1} & \text{else} \end{cases}.$$

Next, we define an equality relation $=_\mathbb{Q}$ on these representatives of the rational numbers.

Definition 1.12. *Equality of rational numbers*
We define the equality between the codes of two rational numbers q_1 and q_2, $=_\mathbb{Q}$, as follows:

$$q_1 =_\mathbb{Q} q_2 :\equiv \frac{\frac{j_1 q_1}{2}}{j_2 q_1 + 1} = \frac{\frac{j_1 q_2}{2}}{j_2 q_2 + 1}$$

if $j_1 q_1$ and $j_1 q_2$ are both even, and analogously in the remaining cases. The equality

$$\frac{a}{b} = \frac{c}{d}$$

is defined to hold iff

$$ad =_0 cb \wedge bd >_0 0.$$

We write $n =_\mathbb{Q} \langle r \rangle$ to express the statement that n represents the rational number r. By using only the representatives of natural numbers and $=_\mathbb{Q}$ we can avoid a formal introduction of the set $\mathbb{Q}$ as the set of the equivalence classes on $\mathbb{N}$ w.r.t. $=_\mathbb{Q}$.
For the embedding $\mathbb{N} \hookrightarrow \mathbb{Q}$, we define for a natural number n its representation as a rational number as follows: $\langle n \rangle = n_\mathbb{Q} := j(2n, 0)$. One can then define the primitive recursive operations $+_\mathbb{Q}, \cdot_\mathbb{Q}$ and the predicates $<_\mathbb{Q}, \leq_\mathbb{Q}$ on $\mathbb{N}$ in WE-HA$^\omega$ such that one can prove (within WE-HA$^\omega$) that

$$(\mathbb{N}, +_\mathbb{Q}, \cdot_\mathbb{Q}, 0_\mathbb{Q}, 1_\mathbb{Q}, <_\mathbb{Q})$$

represents the field of rational numbers, $(\mathbb{Q}, +, \cdot, 0, 1, \leq)$, in WE-HA$^\omega$.

1.4 The Archimedean ordered Field of real Numbers within WE-HA$^\omega$

We represent the real numbers by sequences of rational numbers with a fixed rate of convergence 2^{-n}.

Definition 1.13. *Representation of a real number*
A function $f : \mathbb{N} \mapsto \mathbb{N}$ such that

$$\forall n \, (|fn -_\mathbb{Q} f(n+1)|_\mathbb{Q} <_\mathbb{Q} \langle 2^{-(n+1)} \rangle),$$

which implies

$$\forall n\, \forall k > m \geq n \Big(\, |fm -_{\mathbb{Q}} fk|_{\mathbb{Q}} \leq_{\mathbb{Q}} \sum_{i=m}^{k-1} |fi -_{\mathbb{Q}} f(i+1)|_{\mathbb{Q}} \\ \leq_{\mathbb{Q}} \sum_{i=n}^{\infty} |fi -_{\mathbb{Q}} f(i+1)|_{\mathbb{Q}} <_{\mathbb{Q}} \langle 2^{-n} \rangle \, \Big),$$

represents a real number.

We can arrange for each function f^1 to code a unique real number in this way.

Definition 1.14. Let $f : \mathbb{N} \mapsto \mathbb{N}$ be a function. Define $\widehat{f}$ by

$$\widehat{f}n := \begin{cases} fn & \text{if } \forall k < n \, (|fk -_{\mathbb{Q}} f(k+1)|_{\mathbb{Q}} <_{\mathbb{Q}} \langle 2^{-k-1} \rangle) \\ fk & \text{else} \end{cases},$$

where k is the least number such that $k < n$ and

$$|fk -_{\mathbb{Q}} f(k+1)|_{\mathbb{Q}} \geq_{\mathbb{Q}} \langle 2^{-k-1} \rangle.$$

The function $\widehat{f}$ defines a uniquely determined real number, so we say also that f defines a uniquely determined real number, namely the one represented by $\widehat{f}$.

The functional which maps f to $\widehat{f}$ can be defined primitive recursively in WE-HA$^\omega$. So we can reduce quantifiers ranging over $\mathbb{R}$ to quantifiers ranging over type 1 objects. The usual operations and relations on $\mathbb{R}$ can be defined in WE-HA$^\omega$. As an example we define $=_{\mathbb{R}}$ and $+_{\mathbb{R}}$ as follows:

Definition 1.15. Equality on $\mathbb{R}$ is represented by the equivalence relation $=_{\mathbb{R}}$ of the representatives as follows:

$$f_1^1 =_{\mathbb{R}} f_2^1 :\equiv \forall n (|\widehat{f_1}(n+1) -_{\mathbb{Q}} \widehat{f_2}(n+1)|_{\mathbb{Q}} <_{\mathbb{Q}} \langle 2^{-n} \rangle).$$

We define

$$f_1 <_{\mathbb{R}} f_2 :\equiv \exists n (\widehat{f_2}(n+1) -_{\mathbb{Q}} \widehat{f_1}(n+1) \geq_{\mathbb{Q}} \langle 2^{-n} \rangle),$$

and get $f_1 =_{\mathbb{R}} f_2 \in \Pi_1^0$, $f_1 <_{\mathbb{R}} f_2 \in \Sigma_1^0$, and $f_1 \leq_{\mathbb{R}} f_2 :\equiv \neg(f_2 <_{\mathbb{R}} f_1) \in \Pi_1^0$. The extensionality of these relations can be proved in WE-HA$^\omega$.

Definition 1.16. We define $+_{\mathbb{R}}$ as follows:

$$(f_1 +_{\mathbb{R}} f_2)(k) := \widehat{f_1}(k+1) +_{\mathbb{Q}} \widehat{f_2}(k+1).$$

Analogously for $-_{\mathbb{R}}$.

For the embedding $\mathbb{Q} \hookrightarrow \mathbb{R}$, we define for the coding $n = \langle r \rangle$ of a rational number r its coding $n_{\mathbb{R}} := \lambda k.n_{\mathbb{Q}}$ for the real number corresponding to r. However, usually we omit the intermediate encoding to a rational number and write shortly $r_{\mathbb{R}}$ instead of $(\langle r \rangle)_{\mathbb{R}}$. Again, we do not introduce $\mathbb{R}$ as the set of equivalence classes of representatives, but consider only the representatives themselves. The structure

$$(\mathbb{N}^{\mathbb{N}}, +_{\mathbb{R}}, \cdot_{\mathbb{R}}, 0_{\mathbb{R}}, 1_{\mathbb{R}}, <_{\mathbb{R}})$$

represents the Archimedean ordered field of real numbers $(\mathbb{R}, +, \cdot, 0, 1, <)$ in WE-HA$^\omega$.

1.5 Kuroda's negative Translation

As mentioned earlier, we will use Kuroda's version [91] of Gödel's negative translation [38]. We also give directly the intuitionistically equivalent version for tuples of variables (compare e.g. to [71]).

Definition 1.17 (Kuroda '51)**.** To each formula φ of $\mathcal{L}(\mathsf{WE\text{-}HA}^\omega)$ we associate its *negative translation* φ', which is a formula of the form

$$\varphi' :\equiv \neg\neg\varphi^*,$$

where φ^* is defined inductively as follows:

1. For an atomic formula φ we have:

$$\varphi^* :\equiv \varphi.$$

2. Otherwise, suppose we already know φ^* and ψ^*. We define:

$$\begin{aligned}
(\varphi\wedge\psi)^* &:\equiv \varphi^* \wedge \psi^*,\\
(\varphi\vee\psi)^* &:\equiv \varphi^* \vee \psi^*,\\
(\varphi\to\psi)^* &:\equiv \varphi^* \to \psi^*,\\
(\exists a\ \varphi(a))^* &:\equiv \exists a(\varphi(a))^*,\\
(\forall \underline{a}\ \varphi(\underline{a}))^* &:\equiv \begin{cases} \forall \underline{a}\ \neg\neg(\varphi(\underline{a}))^* & \text{if } \varphi \text{ does not begin with a } \forall\text{-quantifier}\\ \forall \underline{a}\ (\varphi(\underline{a}))^* & \text{else}\end{cases}
\end{aligned}$$

.

This negative translation translates a proof of a formula φ in $\mathsf{WE\text{-}PA}^\omega$ into a proof in $\mathsf{WE\text{-}HA}^\omega$ of φ'.

Theorem 1.18. (See e.g. [93].)
Whenever

$$\mathsf{WE\text{-}PA}^\omega \vdash \varphi$$

we have

$$\mathsf{WE\text{-}HA}^\omega \vdash \varphi'.$$

1.6 Gödel's Functional Interpretation

As mentioned already in the introduction, the D-interpretation was originally developed by Gödel as a possible modification of Hilbert's program. The D-interpretation itself was first presented in a lecture to the Mathematics and Philosophy Clubs at Yale University in 1941 and published in [39]. This first version was for the ordinary Heyting Arithmetic, the version presented here is based on Luckhardt's in [93] and Troelstra's in [125]. Further details on historical background can be found in [7].

Definition 1.19 (Gödel '41)**.** To each formula $\varphi(\underline{a})$ in $\mathcal{L}(\mathsf{WE\text{-}HA}^\omega)$ with the tuple of free variables $\underline{a}$ we associate its *D-interpretation* $\varphi^D(\underline{a})$, which is a formula of the form

$$\varphi^D(\underline{a}) :\equiv \exists \underline{x} \forall \underline{y}\ \varphi_D(\underline{x}, \underline{y}, \underline{a}),$$

where $\varphi_D(\underline{a})$ is a quantifier-free formula. Each of $\underline{x}$ and $\underline{y}$ is a tuple of variables whose types, as well as the length of each tuple, depend only on the logical structure of φ. We

also write $\varphi_D(\underline{x},\underline{y},\underline{a})$ for $\varphi_D(\underline{a})$. If some variables $\underline{z}$ of φ are exhibited, as $\varphi(\underline{z},\underline{a})$, then we write $\varphi_D(\underline{x},\underline{y},\underline{z},\underline{a})$ for $\varphi_D(\underline{a})$.
We define the construction of φ^D inductively as follows

1. For an atomic formula $\varphi(\underline{a})$, the tuples $\underline{x}$ and $\underline{y}$ are empty and we have:

$$\varphi^D(\underline{a}) :\equiv \varphi_D(\underline{a}) :\equiv \varphi(\underline{a}). \tag{0}$$

2. Otherwise, suppose we already know the D-interpretation of $\varphi(\underline{a})$ and $\psi(\underline{b})$ to be as follows:

$$\varphi^D(\underline{a}) \equiv \exists \underline{x} \forall \underline{y}\ \varphi_D(\underline{a}) \quad \text{and} \quad \psi^D(\underline{b}) \equiv \exists \underline{u} \forall \underline{v}\ \psi_D(\underline{b}).$$

where φ_D, ψ_D, $\underline{a}$, $\underline{x}$, $\underline{y}$ and $\underline{b}$ satisfy the assumptions from above and additionally the variable z is not part of any of the tuples above (otherwise we could use a different name for the variable e.g. z_1). We define:

$$(\varphi(\underline{a})\wedge\psi(\underline{b}))^D :\equiv \exists \underline{x},\underline{u} \forall \underline{y},\underline{v} (\varphi_D(\underline{x},\underline{y},\underline{a}) \wedge \psi_D(\underline{u},\underline{v},\underline{b})), \tag{1}$$

$$(\varphi(\underline{a})\vee\psi(\underline{b}))^D :\equiv \exists z^0,\underline{x},\underline{u} \forall \underline{y},\underline{v} \Big((z =_0 0 \to \varphi_D(\underline{x},\underline{y},\underline{a})) \wedge (z \neq_0 0 \to \psi_D(\underline{u},\underline{v},\underline{b})) \Big), \tag{2}$$

$$(\forall z\ \varphi(z,\underline{a}))^D :\equiv \exists \underline{X} \forall z, \underline{y}\ \varphi_D(\underline{X}z,\underline{y},z,\underline{a}), \tag{3}$$

$$(\exists z\ \varphi(z,\underline{a}))^D :\equiv \exists z,\underline{x} \forall \underline{y}\ \varphi_D(\underline{x},\underline{y},z,\underline{a}), \tag{4}$$

$$(\varphi(\underline{a})\to\psi(\underline{b}))^D :\equiv \exists \underline{U},\underline{Y} \forall \underline{x},\underline{v} (\varphi_D(\underline{x},\underline{Y}\,\underline{x}\,\underline{v},\underline{a}) \to \psi_D(\underline{U}\,\underline{x},\underline{v},\underline{b})). \tag{5}$$

The clauses (0),(1) and (4) in the definition 1.19 are very easy to comprehend. The D-interpretation of (3), $(\forall z\ \varphi(z,\underline{a}))^D$, is basically just "Skolemizing" the existentially quantified variables in $\forall z \exists \underline{x} \forall \underline{y} \varphi_D(\underline{x},\underline{y},z,\underline{a})$. The most interesting part of the definition is the D-interpretation of the implication (5). A very detailed and comprehensive explanation can be found in [71].
Originally, Gödel in [39] as well as Avigad and Feferman in [7] defined the D-interpretation of (2):$(\varphi(\underline{a})\vee\psi(\underline{b}))$ in the following intuitionistically equivalent way:

$$\exists z^0,\underline{x},\underline{u} \forall \underline{y},\underline{v} \Big((z=0 \wedge \varphi_D(\underline{x},\underline{y},\underline{a})) \vee (z=1 \wedge \psi_D(\underline{u},\underline{v},\underline{b})) \Big). \tag{+}$$

Here we go after the version used by Troelstra (see e.g. [125]) and Kohlenbach (see e.g. [71]). Apart from that the formula $(\varphi(\underline{a})\vee\psi(\underline{b}))_D$ gets not only quantifier-free but also $\vee$-free this version assures the following idempotency property of the D-interpretation:

Theorem 1.20. *For any formula φ in $\mathcal{L}(\mathsf{WE\text{-}HA}^\omega)$ we have*

$$(\varphi^D)^D \equiv \varphi^D.$$

This would not be the case if we defined (2) in the original way of Gödel. Suppose φ and ψ are atomic formulas. By (+) we have

$$\begin{aligned}
((\varphi\vee\psi)^D)^D &\equiv \Big(\exists z^0 ((z =_0 0 \wedge \varphi) \vee (z =_0 1 \wedge \psi))\Big)^D \\
&\equiv \exists z^0 \Big(\exists z_1^0 \Big((z_1 =_0 0 \wedge (z =_0 0 \wedge \varphi)) \vee (z_1 =_0 1 \wedge (z =_0 1 \wedge \psi)) \Big) \Big) \\
&\not\equiv \exists z^0 ((z =_0 0 \wedge \varphi) \vee (z =_0 1 \wedge \psi)) \equiv (\varphi\vee\psi)^D,
\end{aligned}$$

which obviously contradicts Theorem 1.20.
We should mention that as a consequence $\neg\varphi$ is D-interpreted as (given that $\varphi(\underline{a})$ is D-interpreted as $\exists \underline{x} \forall \underline{y}\ \varphi_D(\underline{a}))$ $\exists \underline{Y} \forall \underline{x} \neg \varphi_D(\underline{x}, \underline{Y}\,\underline{x}, \underline{a})$. This gives us the following "macro" for double negation:

$$(\neg\neg\varphi)^D \equiv \exists \underline{X} \forall \underline{Y} \neg\neg\varphi_D(\underline{X}\,\underline{Y}, \underline{Y}(\underline{X}\,\underline{Y})),$$

which is, provably in WE-HA$^\omega$, equivalent to:

$$\exists \underline{X} \forall \underline{Y} \varphi_D(\underline{X}\,\underline{Y}, \underline{Y}(\underline{X}\,\underline{Y})).$$

Remark 1.21. The treatment of the double negation $(\neg\neg\varphi)^D$ corresponds closely to the so-called no-counterexample interpretation (n.c.i.) due to Kreisel [85]. Both interpretations coincide in the case of $\exists\forall$-formulas (see chapters 2 and 3 below, where this is applied to Cauchy-statements).

We should also mention a partial (weaker) interpretation of the implication, which can be used if the data or information of the premise are not needed for further use of the D-interpretation of a given formula. E.g. if the premise can be proved directly. Typically, we would analyze a proof in which the elimination of an implication is the final step in the proof of the conclusion ψ:

$$\text{WE-HA}^\omega \quad \vdash \quad (\varphi(\underline{a}) \to \psi(\underline{b}))^D \ \to\ \exists \underline{U} \forall \underline{x}, \underline{v}((\forall \underline{y} \varphi_D(\underline{x}, \underline{y}, \underline{a})) \to \psi_D(\underline{U}\,\underline{x}, \underline{v}, \underline{b})).$$

The D-interpretation is sound in the following sense.

Definition 1.22. The *Independence-of-Premise* schema for universal premises: $\text{IP}^\omega_\forall$; is the union (for all types ρ) of

$$\text{IP}^\rho_\forall \quad : \quad (\forall \underline{x} \varphi_{\text{QF}}(\underline{x}) \to \exists y^\rho \psi(y)) \to \exists y^\rho (\forall \underline{x} \varphi_{\text{QF}}(\underline{x}) \to \psi(y)),$$

where y is not free in $\forall \underline{x} \varphi_{\text{QF}}(x)$.

Theorem 1.23. Soundness of D-Interpretation [71]
Let $\mathcal{U}$ be an arbitrary set of purely universal sentences of WE-HA$^\omega$ *and φ a formula of* WE-HA$^\omega$ *containing only $\underline{a}$ as free variables, then*

$$\text{WE-HA}^\omega + \text{AC} + \text{IP}^\omega_\forall + \text{M}^\omega + \mathcal{U} \quad \vdash \quad \varphi(\underline{a})$$

implies that

$$\text{WE-HA}^\omega + \mathcal{U} \quad \vdash \quad \forall \underline{y} \varphi_D(\underline{t}\,\underline{a}, \underline{y}, \underline{a}),$$

where $\underline{t}$ is a suitable tuple of closed terms of WE-HA$^\omega$ *which can be extracted from a given proof.*

In combination with the negative translation as a pre-processing step we get the *ND-interpretation* with the following properties.

Theorem 1.24. *Let $\mathcal{U}$ be an arbitrary set of purely universal sentences of* WE-PA$^\omega$, *φ a formula of* WE-PA$^\omega$ *containing only $\underline{a}$ as free variables and suppose*

$$\text{WE-PA}^\omega \ + \ \text{QF-AC} \ + \ \mathcal{U} \quad \vdash \quad \varphi(\underline{a}),$$

then the ND-interpretation extracts closed terms $\underline{t}$ of WE-HA$^\omega$, *such that*

$$\text{WE-HA}^\omega \ + \ \mathcal{U} \quad \vdash \quad \forall \underline{y} (\varphi')_D(\underline{t}\,\underline{a}, \underline{y}, \underline{a}).$$

Theorem 1.25. Main theorem on program extraction by ND-interpretation [71]
Let $\mathcal{U}$ be an arbitrary set of purely universal sentences of WE-PA$^\omega$ *and $\varphi_{\mathsf{QF}}(\underline{x}, \underline{y})$ be a quantifier-free formula of* WE-PA$^\omega$ *which contains only $\underline{x}^{\underline{\rho}}, \underline{y}^{\underline{\tau}}$ as free variables. Suppose*

$$\mathsf{WE\text{-}PA}^\omega \; + \; \mathsf{QF\text{-}AC} \; + \; \mathcal{U} \quad \vdash \quad \forall \underline{x}^{\underline{\rho}} \exists \underline{y}^{\underline{\tau}} \varphi_{\mathsf{QF}}(\underline{x}, \underline{y}),$$

then the ND-interpretation extracts a tuple of closed terms $\underline{t}$ of WE-HA$^\omega$*, such that*

$$\mathsf{WE\text{-}HA}^\omega \; + \; \mathcal{U} \quad \vdash \quad \forall \underline{x}^{\underline{\rho}} \varphi_{\mathsf{QF}}(\underline{x}, \underline{t}\,\underline{x}).$$

In this thesis, it is almost always the combination of negative translation and functional interpretation that is used. This combination becomes particularly convenient to formulate if one uses a negative translation due to Krivine, as it then coincides with the so-called Shoenfield variant [112] (for the fragment $\{\forall, \vee, \neg\}$) as was shown in [118], where this interpretation is given for the full language (i.e. $\{\forall, \exists, \rightarrow, \vee, \wedge, \neg\}$ with $A \leftrightarrow B :\equiv (A \rightarrow B) \wedge (B \rightarrow A)$).
We now give the definition of this Shoenfield interpretation:

Definition 1.26 ([112, 118]). To each formula $\varphi(\underline{a})$ in $\mathcal{L}(\mathsf{WE\text{-}PA}^\omega)$ with the tuple of free variables $\underline{a}$ we associate its Sh-interpretation $\varphi^{Sh}(\underline{a})$, which is a formula of the form

$$\varphi^{Sh}(\underline{a}) :\equiv \forall \underline{u}\, \exists \underline{x}\, \varphi_{Sh}(\underline{u}, \underline{x}, \underline{a}),$$

where $\varphi_{Sh}(\underline{a})$ is a quantifier-free formula. Each of $\underline{x}$ and $\underline{y}$ is a tuple of variables whose types, as well as the length of each tuple, depend only on the logical structure of φ. We also write $\varphi_{Sh}(\underline{x}, \underline{y}, \underline{a})$ for $\varphi_{Sh}(\underline{a})$. If some variables $\underline{z}$ of φ are exhibited, as $\varphi(\underline{z}, \underline{a})$, then we write $\varphi_{Sh}(\underline{x}, \underline{y}, \underline{z}, \underline{a})$ for $\varphi_{Sh}(\underline{a})$.
We define the construction of φ^{Sh} inductively as follows (with $\underline{x}\,\underline{y}$ denoting $x_1\underline{y}, \ldots, x_n\underline{y}$ for $\underline{x} = x_1, \ldots, x_n$). In the inductive steps we assume that

$$\varphi^{Sh}(\underline{a}) :\equiv \forall \underline{u}\, \exists \underline{x}\, \varphi_{Sh}(\underline{u}, \underline{x}, \underline{a}) \text{ and } \psi^{Sh}(\underline{b}) :\equiv \forall \underline{v}\, \exists \underline{y}\, \psi_{Sh}(\underline{v}, \underline{y}, \underline{b})$$

are already defined.

(S1) $\varphi^{Sh}(\underline{a}) \equiv: \varphi_{Sh}(\underline{a})$ for atomic $\varphi(\underline{a})$,

(S2) $(\neg\varphi)^{Sh} \equiv \forall \underline{f} \exists \underline{u}\, \neg\varphi_{Sh}(\underline{u}, \underline{f}\,\underline{u})$,

(S3) $(\varphi \vee \psi)^{Sh} \equiv \forall \underline{u}, \underline{v} \exists \underline{x}, \underline{y}\, (\varphi_{Sh}(\underline{u}, \underline{x}) \vee \psi_{Sh}(\underline{v}, \underline{y}))$,

(S4) $(\forall z\, \varphi)^{Sh} \equiv \forall z, \underline{u} \exists \underline{x}\, \varphi_{Sh}(z, \underline{u}, \underline{x})$,

(S5) $(\varphi \rightarrow \psi)^{Sh} \equiv \forall \underline{f}, \underline{v} \exists \underline{u}, \underline{y}\, (\varphi_{Sh}(\underline{u}, \underline{f}\,\underline{u}) \rightarrow \psi_{Sh}(\underline{v}, \underline{y}))$,

(S6) $(\exists z\, \varphi)^{Sh} \equiv \forall \underline{U} \exists z, \underline{f}\, \varphi_{Sh}(z, \underline{U}\, z\, \underline{f}, \underline{f}(\underline{U}\, z\, \underline{f}))$,

(S7) $(\varphi \wedge \psi)^{Sh} \leftrightarrow \forall \underline{u}, \underline{v} \exists \underline{x}, \underline{y}\, (\varphi_{Sh}(\underline{u}, \underline{x}) \wedge \psi_{Sh}(\underline{v}, \underline{y}))$.

Remark. The official definition of (S7) in [118] is slightly different from the one given above (where our version is called (S7*)) but is intuitionistically equivalent to that.

As before, a partial (weaker) interpretation of the implication is sufficient when the witnessing data from the premise are not needed for further use of the Sh-interpretation of a given formula. E.g. if the premise can be proved directly. Typically, we would analyze in such cases an implication as:

$$\forall \underline{f} \exists \underline{y} \left(\forall \underline{u}\, \varphi_{Sh}(\underline{u}, \underline{f}\, \underline{u}) \rightarrow \psi_{Sh}(\underline{v}, \underline{y}) \right).$$

Remark. The Shoenfield version of the functional interpretation is often – for obvious reasons – called ∀∃-form, whereas the Dialectica interpretation (and hence also the combination ND of some negative translation N with the Dialectica interpretation D – see above) always is of the form ∃∀. If the Krivine negative translation is used (see [118]), the latter is nothing else but the result of a final application of the axiom schema of quantifier-free choice QF-AC to the Shoenfield interpretation. One should stress though that this passage from the ∀∃-form to the ∃∀-version which also is implicitly present in the soundness theorem of the Shoenfield interpretation (stating the extractability of suitable terms realizing the ∀∃-form) is necessary for the interpretation to be sound for the modus ponens rule.

1.7 Majorizability

We introduce an important structural property of the closed terms of all systems used in this thesis (and more, e.g. all systems used in [71]). We need this property of *majorizability*, which is due to W.A.Howard in [51], to define the third interpretation we will apply. However, there are several applications of majorizability. Mainly, to prove results on growth of the definable functionals of given systems or, more generally, to give bounds on growth or complexity of functionals belonging to a specific class.

Definition 1.27. The relation $x^* \operatorname{maj}_\rho x$ (x^* *majorizes* x) between functionals of type ρ is defined by induction on ρ:

$$x^* \operatorname{maj}_0 x :\equiv x^* \geq_0 x, \tag{1}$$
$$x^* \operatorname{maj}_{\tau\rho} x :\equiv \forall y^*, y (y^* \operatorname{maj}_\rho y \rightarrow x^* y^* \operatorname{maj}_\tau xy). \tag{2}$$

Moreover, Howard showed in [51] the following very useful fact:

Theorem 1.28. W.A.Howard [51]
For each closed term t^ρ of WE-HA$^\omega$ *one can construct a closed term t^* in* WE-HA$^\omega$ *of the same type, such that:*

$$\text{WE-HA}^\omega \quad \vdash \quad t^* \operatorname{maj}_\rho t.$$

1.8 Monotone Functional Interpretation

The monotone functional interpretation, also called *MD-interpretation*, extracts the terms which majorize some functionals realizing the usual functional interpretation directly. It was introduced by Kohlenbach in [62]. However, he studied and used the combination of D-interpretation with majorizability earlier as well, see e.g. [61].
We have the following soundness theorem:

Theorem 1.29. Soundness of MD-interpretation [62],[71]
Let Δ be a set of sentences of the form (φ_{QF} is a quantifier-free formula and the tuple $\underline{r}$ consists of closed terms):

$$\forall \underline{a}^{\underline{\delta}} \exists \underline{b} \leq_{\underline{\sigma}} \underline{r}\, \underline{a} \forall \underline{c}^{\underline{\gamma}} \varphi_{\mathrm{QF}}(\underline{a}, \underline{b}, \underline{c}),$$

and suppose

$$\mathsf{WE\text{-}HA}^{\omega} \ + \ \mathsf{AC} \ + \ \mathsf{IP}^{\omega}_{\forall} \ + \ \mathsf{M}^{\omega} \ + \ \Delta \ \vdash \ \varphi(\underline{a}),$$

then MD-interpretation extracts closed terms $\underline{t}^*$ of WE-HA$^{\omega}$ *such that*

$$\mathsf{WE\text{-}HA}^{\omega} \ + \ \tilde{\Delta} \ \vdash \ \exists \underline{x}(t^* \,\mathrm{maj}\, x \wedge \forall \underline{a}, \underline{y} \varphi_D(\underline{x}\,\underline{a}, \underline{y}, \underline{a})),$$

where $\tilde{\Delta}$ is the corresponding set of the Skolem normal forms of sentences in Δ:

$$\tilde{\Delta} := \left\{ \tilde{\varphi} :\equiv \exists \underline{B} \le \underline{r} \forall \underline{a}, \underline{c} \varphi_{\mathsf{QF}}(\underline{a}, \underline{B}\,\underline{a}, \underline{c}) \ : \ \varphi \equiv \forall \underline{a}^{\underline{\delta}} \exists \underline{b} \le_{\sigma} \underline{r}\,\underline{a} \forall \underline{c}^{\underline{\gamma}} \varphi_{\mathsf{QF}}(\underline{a}, \underline{b}, \underline{c}) \in \Delta \right\}.$$

Definition 1.30. Considering the theorem 1.29, we say that the tuple of terms $\underline{t}^*$ satisfies the *monotone functional interpretation*, or simply *MD-interpretation*, of φ.

Kohlenbach connects in [71] the MD-interpretation with proof mining by the following theorem:

Theorem 1.31. Main theorem on uniform bound extraction by MD-interpretation [71]
Let Δ be as above and $\varphi(x^1, y^{\rho}, z^{\tau})$ an arbitrary formula containing only x,y,z free. Let $s^{\rho(1)}$ be a closed term of WE-HA$^{\omega}$ *and the type level of τ not greater than* 2.
Then if one can prove:

$$\mathsf{WE\text{-}HA}^{\omega} + \mathsf{AC} + \mathsf{IP}^{\omega}_{\forall} + \mathsf{M}^{\omega} + \Delta \ \vdash \ \forall x^1 \forall y \le_{\rho} sx \exists z^{\tau} \varphi(x, y, z)$$

one can also extract a closed term t of WE-HA$^{\omega}$ *s.t.*

$$\mathsf{WE\text{-}HA}^{\omega} + \mathsf{AC} + \mathsf{IP}^{\omega}_{\forall} + \mathsf{M}^{\omega} + \Delta \ \vdash \ \forall x^1 \forall y \le_{\rho} sx \exists z^{\tau} \le_{\tau} tx \ \varphi(x, y, z).$$

As in the case of D-interpretation, we can combine the MD-interpretation with the negative translation to obtain the *NMD-interpretation*. In this case the main theorem, again due to Kohlenbach (see e.g. [71]), becomes:

Theorem 1.32. Main theorem on uniform bound extraction by NMD-interpretation [71]
Let Δ,$\tilde{\Delta}$ be as above and $\varphi_{\mathsf{QF}}(x^1, y^{\rho}, z^{\tau})$ be a quantifier-free formula of WE-PA$^{\omega}$ *containing only x,y,z free. Let $s^{\rho(1)}$ be a closed term of* WE-HA$^{\omega}$ *and the type level of τ not greater than* 2.
Then if one can prove:

$$\mathsf{WE\text{-}PA}^{\omega} + \mathsf{QF\text{-}AC} + \Delta \ \vdash \ \forall x^1 \forall y \le_{\rho} sx \exists z^{\tau} \varphi_{\mathsf{QF}}(x, y, z)$$

one can also extract a closed term t of WE-HA$^{\omega}$ *s.t.*

$$\mathsf{WE\text{-}HA}^{\omega} + \tilde{\Delta} \ \vdash \ \forall x^1 \forall y \le_{\rho} sx \exists z^{\tau} \le_{\tau} tx \ \varphi_{\mathsf{QF}}(x, y, z).$$

Both theorems, 1.31 and 1.32, apply also to tuples of variables, where tuples of terms are extracted. The proofs are given e.g. in [71].

1.9 A specific metatheorem

Metatheorems which guarantee that specific uniform bounds for the realizers can be extracted were developed in [70] and [37] (see also [71]) and are applicable to many theorems concerning a wide range of classes of maps and abstract spaces. For example, they were successfully applied to the ergodic theorems we mention in Figure 1 in the chapter on Wittmann's strong non-linear ergodic theorem or to asymptotic regularity theorems in metric fixed point theory [81]. In the last mentioned example, the authors infer from the metatheorems that uniform bounds exist and derive them explicitly.
To apply the metatheorems, one needs the analyzed theorem to meet only two conditions:

1. The proof does not use axioms or rules which are too strong.

2. The analyzed theorem in its logical form is not too complex in terms of quantification.

To formalize the first condition we start with a logical system for so called full classical analysis introduced by Spector in [117].[3] Kohlenbach extended this system by an additional basic type and its defining axioms representing a given abstract space and its properties. Kohlenbach also considers cases, where a specific subset of such a space (or rather its characteristic function) has to exist as a constant. For instance in [71] Kohlenbach defines such systems for the theory of metric, hyperbolic, normed, uniformly convex or Hilbert spaces – if required – together with a (bounded) convex subset.[4] For the purpose of this thesis (i.e. simply for a pre-Hilbert space) this extended system is denoted by $\mathcal{A}^\omega[X, \langle\cdot,\cdot\rangle]$. In general, the system can be extended to arbitrary Hilbert spaces, however it turns out that the completeness is not necessary for the proof that we analyze in the corresponding chapter.
The second condition has to be investigated for each theorem specifically, depending on the given theorem and the metatheorem we wish to use. Examples are metastable versions of formulas expressing the convergence or fixed point properties of nonexpansive, Lipschitz, weakly quasi-nonexpansive or uniformly continuous functions even simply functions which are majorizable (see Corollary 6.6 in [37] and Theorem 1.33 below). The metatheorem applicable in the scenario we discuss in chapter 3 follows from Corollary 6.6.7) in [37]. In particular with the theory $\mathcal{A}^\omega[X, \langle\cdot,\cdot\rangle]$ with an additional parameter for an arbitrary subset S of the abstract Hilbert space X.[5]

Theorem 1.33 (Gerhardy-Kohlenbach [37] - specific case 1). *Let $\varphi_\forall$, resp. $\psi_\exists$, be $\forall$- resp. $\exists$-formulas that contain only x, z, f free, resp. x, z, f, v free. Assume that $\mathcal{A}^\omega[X, \langle\cdot,\cdot\rangle, S]$ proves the following sentence:*

$$\forall x \in \mathbb{N}^{\mathbb{N}}, z \in S, f \in S^S \left(\varphi_\forall(x,z,f) \to \exists v \in \mathbb{N}\ \psi_\exists(x,z,f,v)\right).$$

Then there is a computable functional $F : \mathbb{N}^{\mathbb{N}} \times \mathbb{N} \times \mathbb{N}^{\mathbb{N}} \to \mathbb{N}$ s. t. the following holds in

[3] In particular this system covers full comprehension over numbers, including also full second order arithmetic.

[4] In any such abstract space, its metric plays a major role as two objects are defined to be equal, if and only if their distance is zero.

[5] This is analogous to the case where we add C to the theory for normed space, but this time without any additional axioms.

all non-trivial (real) inner product spaces $(X, \langle\cdot,\cdot\rangle)$ and for any subset $S \subseteq X$

$$\begin{aligned}&\forall x \in \mathbb{N}^{\mathbb{N}}, z \in S, b \in \mathbb{N}, f \in S^S, f^* \in \mathbb{N}^{\mathbb{N}}\\ &\big(\mathrm{Maj}(f^*, f) \wedge \|z\| \leq b \wedge \varphi_\forall(x,z,f) \to \exists v \leq F(x,b,f^*)\ \psi_\exists(x,z,f,v)\big),\end{aligned}$$

where

$$\mathrm{Maj}(f^*, f) :\equiv \forall n \in \mathbb{N} \forall z \in S\big(\|z\| \leq_{\mathbb{R}} n \to \|f(z)\| \leq_{\mathbb{R}} f^*(n)\big).$$

The theorem holds analogously for finite tuples.

1.10 Models of E-PA$^\omega$

We will introduce three models of E-PA$^\omega$:

- The full set-theoretic model: $\mathcal{S}^\omega$.
- The model of all sequentially continuous functionals: $\mathcal{C}^\omega$.
- The model of all strongly majorizable functionals: $\mathcal{M}^\omega$.

Note that as models of E-PA$^\omega$ these are also models of any other system we presented so far.

Definition 1.34. The *full set-theoretic model* is defined inductively as the type-structure of all set-theoretic functionals:

$$\begin{aligned}\mathcal{S}_0 &:= \mathbb{N}, &(1)\\ \mathcal{S}_{\tau\rho} &:= \{\text{ all set-theoretic functionals } F : \mathcal{S}_\rho \to \mathcal{S}_\tau\}, &(2)\\ \mathcal{S}^\omega &:= \bigcup_{\rho \in \boldsymbol{T}} \mathcal{S}_\rho. &(3)\end{aligned}$$

To define $\mathcal{C}^\omega$ we need the following definition first:

Definition 1.35. Let $(X, \to_X)$ and $(Y, \to_Y)$ be two L-spaces. A function $f : X \to Y$ is called continuous if $f(p_n) \to_Y f(p)$ whenever $p_n \to_X p$. We denote the set of all continuous functions from X to Y by $\mathcal{C}(X,Y)$.
Here, by L-space we mean a 'limit space' as defined in [90].

Definition 1.36. [109] The *model of all sequentially continuous functionals* is defined inductively as follows:

$$\begin{aligned}&\mathcal{C}_0 := \mathbb{N}, && p_n \to_0 p :\equiv \exists k \forall m > k\ p_m = p &(1)\\ &\mathcal{C}_{\tau\rho} := \mathcal{C}(\mathcal{C}_\rho, \mathcal{C}_\tau), && f_n \to_{\tau\rho} f :\equiv \forall (p_n) \in \mathcal{C}_\rho^{\mathbb{N}}, p \in \mathcal{C}_\rho\ \big(p_n \to_\rho p \Rightarrow f_n(p_n) \to_\tau f(p)\big) &(2)\\ &\mathcal{C}^\omega := \bigcup_{\rho \in \boldsymbol{T}} \mathcal{C}_\rho. && &(3)\end{aligned}$$

To define $\mathcal{M}^\omega$ we will make use of the variant s-maj of Howard's majorization relation maj.

Definition 1.37. **[15]** The *model of strongly majorizable functionals* is the type structure $\mathcal{M}^\omega$ defined inductively as follows:

$$\mathcal{M}_0 := \mathbb{N},$$
$$n \,\text{s-maj}_0\, m \;:\equiv\; n \geq m \wedge n, m \in \mathbb{N} \tag{1}$$
$$\mathcal{M}_{\tau(\rho)} := \left\{ x \in \mathcal{M}_\tau^{\mathcal{M}_\rho} \;:\; \exists x^* \in \mathcal{M}_\tau^{\mathcal{M}_\rho} \, (x^* \,\text{s-maj}_{\tau(\rho)}\, x) \right\},$$
$$x^* \,\text{s-maj}_{\tau(\rho)}\, x \;:\equiv\; \forall y^*, y \in \mathcal{M}_\rho \left(y^* \,\text{s-maj}_\rho\, y \;\rightarrow\; x^* y^* \,\text{s-maj}_\tau\, x^* y, xy \right) \wedge x^*, x \in \mathcal{M}_\tau^{\mathcal{M}_\rho} \tag{2}$$
$$\mathcal{M}^\omega := \bigcup_{\rho \in \boldsymbol{T}} \mathcal{M}_\rho, \tag{3}$$

where $\mathcal{M}_\tau^{\mathcal{M}_\rho}$ is the set of all total set-theoretic mappings from $\mathcal{M}_\rho$ into $\mathcal{M}_\tau$.

Remark 1.38. While $\mathcal{M}^\omega$ and $\mathcal{C}^\omega$ are models of $\mathsf{E}\text{-}\mathsf{PA}^\omega$ with bar recursion (see definition 1.39) added, $\mathcal{S}^\omega$ is obviously not. However, the models start to differ only from type 2 on, where we still have: $\mathcal{C}_2 \subset \mathcal{M}_2 \subset \mathcal{S}_2$. So, if we use bar recursion and Peano Arithmetic to produce a functional F of type level 2, we know that it is a well defined functional in $\mathcal{C}^\omega$ therefore also in $\mathcal{M}^\omega$ and in $\mathcal{S}^\omega$ and so a total function: $\mathbb{N}^\mathbb{N} \mapsto \mathbb{N}$.

1.11 Bar Recursion

We give the definition in the form presented by Spector in [117]. Note that this form already uses the fact that tuples of variables of any type can be contracted into single variables. Alternatively, one could use a simultaneous form of bar recursion (see [71]).

Definition 1.39. The *bar recursor* $\mathsf{B}_{\rho,\tau}$ is defined by:

$$\mathsf{B}_{\rho,\tau} yzunx :=_\tau \begin{cases} zn(\overline{x,n}) & \text{if } y(\overline{x,n}) <_0 n \\ u(\lambda D^\rho . \mathsf{B}_{\rho,\tau} yzu(n+1)(\overline{x,n} * D))n(\overline{x,n}) & \text{else} \end{cases},$$

where

$$(\overline{x,n})(k^0) =_\rho \begin{cases} x(k) & \text{if } k <_0 n \\ 0^\rho & \text{else} \end{cases}$$

and

$$(\overline{x,n} * D)(k^0) =_\rho \begin{cases} x(k) & \text{if } k <_0 n \\ D & \text{if } k =_0 n \,. \\ 0^\rho & \text{else} \end{cases}$$

Remark 1.40. Note that $\overline{f,n}$, which is still a type 1 function, is not the same as $\bar{f}n$, what is a type 0 object (see the section Introduction, paragraph Notation and Common Expressions).

For better readability Spector presented in [117] the recursor Φ,

$$\Phi_\rho yunxm :=_\rho \begin{cases} xm, \text{ if } m <_0 n \\ 0^\rho, \text{ if } m \geq_0 n \wedge y(\overline{x,n}) < n \\ \Phi_\rho yun'(\overline{x,n} * D_0)m, \text{ otherwise} \end{cases} \qquad D_0 =_\rho un(\lambda D^\rho . \Phi_\rho yun'(\overline{x,n} * D)),$$

for a special form of bar recursion (it was shown that, with appropriate type adjustments, Spector's variant is equivalent to the standard version in [101]). As in this thesis we deal only with arithmetical comprehension over numbers we don't need the bar recursion for all types. The recursor $\mathsf{B}_{0,1}$ is fully sufficient and corresponds to Φ_0 which can be defined primitive recursively in $\mathsf{B}_{0,1}$ as follows:

Definition 1.41.

$$\Phi_0 yunx =_1 \mathsf{B}_{0,1} yz\tilde{u}nx,$$

where

$$znxm :=_0 \begin{cases} xm & \text{if } m <_0 n \\ 0^0 & \text{else} \end{cases},$$

$$\tilde{u}vnxm :=_0 \begin{cases} xm & \text{if } m <_0 n \\ v(unv)m & \text{else} \end{cases}.$$

However, for any practical issue we will use the direct definition:

$$\Phi_0 yunxm :=_0 \begin{cases} xm, \text{ if } m <_0 n \\ 0^0, \text{ if } m \geq_0 n \wedge y(\overline{x,n}) < n \\ \Phi_0 yun'(\overline{x,n} * D_0)m, \text{ otherwise} \end{cases},$$

where

$$D_0 =_0 un(\lambda D^0.\Phi_0 yun'(\overline{x,n} * D)).$$

Furthermore, to be able to properly analyze the complexity of the witnessing functionals in later sections, we introduce Howard's schemas of restricted bar recursion as given in [53].

Definition 1.42. The *restricted bar recursor* for *Scheme 1*, $\mathsf{B'}_1$, is defined by:

$$\mathsf{B'}_1 y^2 z^{(2)0} un^0 x^1 :=_0$$
$$\begin{cases} zn(\overline{x,n}) & \text{if } y(\overline{x,n}) <_0 n \\ u(\mathsf{B'}_1 yzu(n+1)(\overline{x,n}*0))(\mathsf{B'}_1 yzu(n+1)(\overline{x,n}*1))n(\overline{x,n}) & \text{else} \end{cases},$$

and for *Scheme 2*, $\mathsf{B'}_2$, by:

$$\mathsf{B'}_2 y^2 u^1 n^0 x^1 :=_0 \begin{cases} 0 & \text{if } y(\overline{x,n}) <_0 n \\ 1 + \mathsf{B'}_2 yu(n+1)(\overline{x,n} * (un)) & \text{else} \end{cases}.$$

Remark 1.43. Note that, $\mathsf{B'}_1$ and $\mathsf{B'}_2$ are just special forms of $\mathsf{B}_{0,0}$ since we have:

$$\mathsf{B'}_1 y^2 z^{(2)0} un^0 x^1 = \mathsf{B}_{0,0} yz\tilde{u}nx,$$

by setting

$$\tilde{u}vnx := u(v0)(v1)nx$$

i.e. it holds

$$\tilde{u}(\lambda D^0.\mathsf{B}_{0,0} yz\tilde{u}(n+1)(\overline{x,n} * D))n(\overline{x,n}) =$$
$$u\Big((\lambda D^0.\mathsf{B}_{0,0} yz\tilde{u}(n+1)(\overline{x,n}*D))0\Big)\Big((\lambda D^0.\mathsf{B}_{0,0} yz\tilde{u}(n+1)(\overline{x,n}*D))1\Big)n(\overline{x,n}).$$

And also:

$$\mathsf{B'}_2 y^2 u^1 n^0 x^1 = \mathsf{B}_{0,0} y \mathbf{0}^{(2)0} \tilde{u} n x,$$

by setting

$$\tilde{u} v n x := 1 + v(un)$$

i.e. it holds

$$\tilde{u}\big(\lambda D^0.\mathsf{B}_{0,0} y z \tilde{u}(n+1)(\overline{x,n} * D)\big) n(\overline{x,n}) = \\ 1 + \big(\lambda D^0.\mathsf{B}_{0,0} y z \tilde{u}(n+1)(\overline{x,n} * D)\big)(un).$$

Furthermore applied only to 0,1-sequences (i.e. type one functions $x : \mathbb{N} \mapsto \{0,1\}$), as is always the case in this thesis, is $\mathsf{B'}_2$ a special form of $\mathsf{B'}_1$ since we have:

$$\mathsf{B'}_2 y^2 u^1 n^0 x^1 = \mathsf{B'}_1 y^2 \mathbf{0}^{(2)0} \tilde{u} n^0 x^1,$$

by setting

$$\tilde{u} v_1 v_2 n x := \begin{cases} 1 + v_1 & \text{if } un =_0 0 \\ 1 + v_2 & \text{else} \end{cases}.$$

i.e. it holds

$$\tilde{u}\big(\mathsf{B'}_1 y z \tilde{u}(n+1)(\overline{x,n} * 0)\big)\big(\mathsf{B'}_1 y z \tilde{u}(n+1)(\overline{x,n} * 1)\big) n(\overline{x,n}) = \\ \begin{cases} 1 + \big(\mathsf{B'}_1 y z \tilde{u}(n+1)(\overline{x,n} * 0)\big) & \text{if } un =_0 0 \\ 1 + \big(\mathsf{B'}_1 y z \tilde{u}(n+1)(\overline{x,n} * 1)\big) & \text{else} \end{cases}.$$

1.12 Law of excluded Middle

The law of excluded middle over numbers for Σ^0_1-formulas is defined as follows:

$$\Sigma^0_1\text{-LEM}(f) \quad : \quad \forall x \exists y \forall z \, \big(f(x,y) =_0 0 \vee f(x,z) \neq_0 0\big).$$

Remark 1.44. Usually this law would be formulated in the following, under $\mathsf{WE\text{-}HA}^\omega$ equivalent, way:

$$\forall x \, \big(\exists y \, f(x,y) =_0 0 \vee \forall z \, f(x,z) \neq_0 0\big).$$

To obtain the ND-interpretation we first use the negative translation $(\Sigma^0_1\text{-LEM}(f))'$:

$$\neg\neg\forall x^0 \neg\neg\exists y^0 \forall z^0 \, \neg\neg\big(f(x,y) =_0 0 \vee f(x,z) \neq_0 0\big),$$

which is intuitionistically equivalent to

$$\forall x \neg\neg\exists y \forall z \, (f(x,y) =_0 0 \vee f(x,z) \neq_0 0).$$

Now, following very strictly all rules, we get the functional interpretation of $(\Sigma^0_1\text{-LEM})'$:

$$\forall Z^{10}, f^{10}, x^0 \\ \Big(\big(t_D(Z(t_Y Zxf)(t_D Zxf))Zxf =_0 0 \wedge f(x, t_Y Zxf) =_0 0\big) \vee \\ \big(t_D(Z(t_Y Zxf)(t_D Zxf))Zxf =_0 1 \wedge f(x, Z(t_D Zxf)(t_Y Zxf)) \neq_0 0\big) \Big),$$

where

$$t_D(n^0, Z, x, f) :=_0 \begin{cases} 1 & \text{if } f(x, Z11) \neq_0 0 \\ 0 & \text{else} \end{cases},$$

$$t_Y(Z, x, f) :=_0 \begin{cases} 1 & \text{if } f(x, Z11) \neq_0 0 \\ Z11 & \text{else} \end{cases}.$$

However, we can see that this interpretation is unnecessarily blown up (e.g. the parameter n corresponding to $Z(t_Y Zxf)(t_D Zxf)$ in the term t_D can be completely ignored). Therefore, we give a more readable, but still intuitionistically equivalent, form:

$$\forall Z^1, f^{10}, x^0 \left(f(x, t_Y Zxf) =_0 0 \vee f(x, Z(t_Y Zxf)) \neq_0 0 \right),$$

where

$$t_Y(Z, x, f) :=_0 \begin{cases} 1 & \text{if } f(x, Z1) \neq_0 0 \\ Z1 & \text{else} \end{cases}.$$

Remark 1.45. From now on we will always simply prefer better readability whenever possible.

Looking closely at the case distinction of $f(x, Z1) \neq_0 0$ being true or not, it is easy to see the correctness of the interpretation.

1.13 Double Negation Shift

The Double Negation Shift schema over numbers for Σ^0_2 formulas is defined as follows:

$$\Sigma^0_2\text{-DNS}(f) \quad : \quad \forall x^0 \neg\neg\exists y^0 \forall z^0\, f(x, y, z) =_0 0 \;\rightarrow\; \neg\neg\forall x^0 \exists y^0 \forall z^0\, f(x, y, z) =_0 0.$$

This schema is interpreted under the functional interpretation using Bar Recursion essentially in the same way as full DNS as presented by Spector in [117]:

$$\forall A^{2(0)}, T^2, W^2 \left(f(t^0_x, At_x t^1_B, t_B(At_x t_B)) =_0 0 \rightarrow f(Tt^1_U, t_U(Tt_U), Wt_U) =_0 0 \right)$$

$$t_x :=_0 Tx^1_0, \quad t_U :=_1 x_0, \quad t_B =_1 \lambda D^0.W^2(E^{1(0)}_{Tx_0} D),$$

where

$$x_0 :=_1 \Phi_0 Tu^{1(1(0))} 0^0 0^1$$

$$E_{n^0} :=_{1(0)} \lambda D^0.\Phi_0 Tun'(\overline{x_0, n} * D)$$

$$un^0 v^{1(0)} :=_0 An \left(\lambda D^0.W(v(D)) \right).$$

Remark 1.46. The detailed steps of the interpretation show that any term can depend on all three variables A, T and W. However, for better readability we present the result in its simpler form. E.g. by t_x we mean in fact $t_x ATW$. To be absolutely correct we would have to write: $t_x :=_{((0(20))(2))(2)} \lambda A, T, W.Tx_0^{(2)2} AW$.

The correctness proof as well as some more details can be found e.g. in [71].

1.14 Axiom of Choice

The very well known Axiom of Choice is used in a wide variety of forms in Proof Theory. We follow Troelstra (see [125]) and define:

Definition 1.47. Axiom of Choice

$$
\begin{aligned}
\mathsf{AC}^{\rho,\tau} &: \quad \forall x^\rho \exists y^\tau \varphi(x,y) \to \exists Y^{\tau(\rho)} \forall x^\rho \varphi(x, Zx) \\
\mathsf{AC} &: \quad \bigcup_{\rho,\tau \in \boldsymbol{T}} \mathsf{AC}^{\rho,\tau} \\
\mathsf{QF\text{-}AC}^{\rho,\tau} &: \quad \forall x^\rho \exists y^\tau \varphi_{\mathsf{QF}}(x,y) \to \exists Y^{\tau(\rho)} \forall x^\rho \varphi_{\mathsf{QF}}(x, Zx) \\
\mathsf{QF\text{-}AC} &: \quad \bigcup_{\rho,\tau \in \boldsymbol{T}} \mathsf{QF\text{-}AC}^{\rho,\tau},
\end{aligned}
$$

where φ_{QF} is quantifier-free.

Moreover, we define an arithmetical version for universal formulas:

Definition 1.48. Axiom of Choice over numbers for Π^0_1 formulas, Π^0_1-$\mathsf{AC}(f)$:

$$\forall x^0 \exists y^0 \forall z^0\, f(x,y,z) =_0 0 \;\; \to \;\; \exists g^1 \forall x^0, z^0\, f(x, gx, z) =_0 0).$$

This axiom is interpreted under the functional interpretation as:

$$\forall x^0, z^0, Y^1 \Big(f(t_X Yxz, Y(t_X Yxz), t_Z Yxz) =_0 0 \to f\big(x, (t_G Y)x, z\big) =_0 0 \Big),$$

$$t_X(Y^1, x^0, z^0) :=_0 x, \quad t_Z(Y^1, x^0, z^0) :=_0 z, \quad t_G(Y^1) :=_1 Y.$$

We can see the correctness immediately. For further use we will need also the functional interpretation of the double negation of Π^0_1-AC, $\neg\neg\Pi^0_1$-AC, which is also straightforward.

$\neg\neg\Pi^0_1$-$\mathsf{AC}(f)$:

$$\neg\neg(\forall x^0 \exists y^0 \forall z^0\, f(x,y,z) =_0 0 \;\; \to \;\; \exists g^1 \forall u^0, v^0\, f(u, gu, v) =_0 0)).$$

This double negation is interpreted under the functional interpretation as:

$$
\begin{aligned}
\forall Y^1, U, V \Big(\; & f(t_X YUV, Y(t_X YUV), t_Z YUV) =_0 0 \\
& \to f(U(t_G Y), (t_G Y)(U(t_G Y)), V(T_G Y)) =_0 0 \; \Big),
\end{aligned}
$$

where

$$
\begin{aligned}
t_X(Y, U, V) &:=_0 UY, \\
t_Z(Y, U, V) &:=_0 VY, \\
t_G(Y) &:=_1 Y.
\end{aligned}
$$

To prove $(\Pi^0_1$-$\mathsf{AC}(f))'$,

$$\forall x^0 \neg\neg \exists y^0 \forall z^0\, f(x,y,z) =_0 0 \;\; \to \;\; \neg\neg \exists g^1 \forall x^0, z^0\, f(x, gx, z) =_0 0,$$

we need only DNS and $\neg\neg(\Pi^0_1\text{-AC}(f))$. The double negation $\neg\neg(\Pi^0_1\text{-AC}(f))$ is intuitionistically equivalent to:

$$\neg\neg\forall x^0\exists y^0\forall z^0\, f(x,y,z) =_0 0 \;\to\; \neg\neg\exists g^1\forall x^0, z^0\, f(x,gx,z) =_0 0. \qquad (+)$$

Using $(+)$ we can prove $(\Pi^0_1\text{-AC}(f))'$ by a single use of the rule

$$\frac{\varphi \to \psi \quad \psi \to \chi}{\varphi \to \chi},$$

since by setting

$$\varphi :\equiv \forall x^0 \neg\neg\exists y^0\forall z^0\, f(x,y,z) =_0 0,$$
$$\psi :\equiv \neg\neg\forall x^0\exists y^0\forall z^0\, f(x,y,z) =_0 0,$$
$$\chi :\equiv \neg\neg\exists g^1\forall x^0, z^0\, f(x,gx,z) =_0 0,$$

we get

$$\frac{\Sigma^0_2\text{-DNS}(f) \quad \neg\neg(\Pi^0_1\text{-AC}(f))}{(\Pi^0_1\text{-AC}(f))'}.$$

To find the interpreting terms we follow the term construction as given e.g. in [71]. We will use the interpretations given above (see section 1.13 for the interpretation of DNS) and solve the following equations:

$$t_X YUV =_0 Tt_U$$
$$Y(t_X YUV) =_0 t_U(Tt_U)$$
$$t_Z YUV =_0 Wt_U.$$

Now, easily by setting $Y := t_U$ we get $T = U$ and $V = W$, which leads us to precisely the same interpretation as the one of DNS:

$$\forall A^{2(0)}, T^2, W^2 \left(f(t_x, At_xt_B, t_B(At_xt_B)) =_0 0 \to f(Tt_U, t_U(Tt_U), Wt_U) =_0 0\right),$$

where t_x, t_B and t_U are as in section 1.13.

Remark 1.49. We could conclude this already from the fact that apart from DNS the only used principle was a double negation of intuitionistically very weak and trivially interpretable axiom Π^0_1-AC.

1.15 Arithmetical Comprehension

The *Schema of Comprehension* is known in several forms. We give following four in $\mathcal{L}(\mathsf{WE\text{-}HA}^\omega)$.

Definition 1.50. The schema:

$$\exists f^{0(\underline{\tau})} \forall \underline{x}^{\underline{\tau}} \; (\, \varphi(\underline{x}) \leftrightarrow f\underline{x} =_0 0 \,)$$

is called

- *Full Comprehension*, abbreviation: CA; iff there are no restrictions on the formula φ.
- *Arithmetical Comprehension*, abbreviations: Π^0_∞-CA, ACA or CA_{ar}; iff all quantifiers in φ range over numbers only. Equivalently, we may say φ is an arithmetical formula or φ is a Π^0_∞ formula.
- (Arithmetical) *Comprehension over numbers*, abbreviations: $(\Pi^0_\infty$-$)\mathsf{CA}^0$, $(\mathsf{A})\mathsf{CA}^0$ or $\mathsf{CA}^0_{(ar)}$; iff x is a single number, i.e. $\underline{\tau} = (0)$.
- *Arithmetical Comprehension over numbers for purely universal formulas*, Π^0_1-CA, iff φ is a Π^0_1 formula and x is a number. Using the fact that in all systems we use, we have characteristic functions for quantifier-free formulas, see e.g. [71], we obtain the following equivalent formulation:

$$\Pi^0_1\text{-CA} \quad : \quad \forall f^{0(00)} \underbrace{\exists g^1 \forall x^0 \Big(\, (\forall y^0 \, f(x,y) =_0 0) \leftrightarrow gx =_0 0 \, \Big)}_{\equiv : \Pi^0_1\text{-CA}(f^{0(00)})}.$$

Remark 1.51. Since we can derive any instance of CA^0_{ar} in $\mathsf{WE\text{-}PA}^\omega$ by iterated application of Π^0_1-CA, the last two versions are equivalent. However, this is only the case for full Π^0_1-CA and not for concrete instances Π^0_1-CA(f).

The schema of arithmetical comprehension is one of the core schemas we need for the interpretation of proofs based on sequential compactness. It is also considered to be essential for formalizing large parts of classical analysis, see [114]. For our purposes the suitable instances of the schema of comprehension over numbers for Π^0_1 (or equivalently for Σ^0_1) formulas are fully sufficient. As in the case of Π^0_1-AC the whole complexity of the interpretation is given by Σ^0_2-DNS with a primitive recursive case distinction from Σ^0_1-LEM.
The even in $\mathsf{E\text{-}PA}^\omega$ stronger, but using $\mathsf{QF\text{-}AC}^{0,0}$ already in $\widehat{\mathsf{WE\text{-}HA}}^\omega\!\restriction$ to Π^0_1-CA equivalent, form called modified schema of comprehension over numbers for Π^0_1 formulas,

$$\Pi^0_1\text{-}\widehat{\mathsf{CA}}(f) \quad : \quad \exists g^1 \forall x^0, z^0 (fx(gx) =_0 0 \vee fxz \neq_0 0),$$

is very practical for finding the functional interpretation as well as for finding the functional interpretation of the negative translation.

Remark 1.52. To see the fact that Π^0_1-CA is weaker than Π^0_1-$\widehat{\mathsf{CA}}$ observe that Π^0_1-$\widehat{\mathsf{CA}}$ proves the existence of functions growing faster than any function in $\mathcal{T}$, which Π^0_1-CA

does not, as the function f in Π^0_1-CA is always majorized by $\mathbf{1}^1$.
On the other hand, for suitable primitive recursive functional F, we can prove Π^0_1-CA(Ff) implies Π^0_1-$\widehat{\mathsf{CA}}(f)$ in $\widehat{\mathsf{WE\text{-}HA}}^{\omega}\!\restriction + \mathsf{QF\text{-}AC}^{0,0}$ in a similar way as we prove $(\Pi^0_1\text{-}\mathsf{CA}(f))'$ below. First, we apply Π^0_1-CA to the intuitionistically valid law Σ^0_1-LEM(f):

$$\underbrace{\forall x \exists y \forall z \quad \underbrace{(f(x,y) =_0 0 \vee f(x,z) \neq_0 0)}_{:\equiv F(f,x,y,z)=_0 0 \text{ for a suitable prim. rek. F}}}_{\text{apply } \Pi^0_1\text{-CA}(Ff)},$$

obtaining a $g^{0(00)}$ s. t. $\forall x^0 \exists y^0\ (gxy =_0 0 \leftrightarrow \forall z^0\ Ffxyz =_0 0)$. By $\mathsf{QF\text{-}AC}^{0,0}$ we get a function h^1 such that it holds: $\forall x\ g(x, hx) =_0 0$, so finally we obtain Π^0_1-$\widehat{\mathsf{CA}}(f)$.
On the other hand, given Π^0_1-$\widehat{\mathsf{CA}}(f)$ we can define a new g^1 as $\lambda x^0.f^{0(00)}(x, g^1x)$ obtaining Π^0_1-CA(f).

The negative translation of the modified schema of comprehension over numbers for Π^0_1 formulas,

$$(\Pi^0_1\text{-}\widehat{\mathsf{CA}}(f))' \quad : \quad \neg\neg\exists g^1 \forall x^0, z^0 (fx(gx) =_0 0 \vee fxz \neq_0 0),$$

follows directly by applying $(\Pi^0_1\text{-}\mathsf{AC}(G))'$ to $(\Sigma^0_1\text{-}\mathsf{LEM}(f))'$ for a suitable φ.
By defining

$$G_f(x,y,z) :\equiv \begin{cases} 0 & \text{if } f(x,z) \neq_0 0 \\ f(x,y) & \text{else} \end{cases},$$

$(\Sigma^0_1\text{-}\mathsf{LEM}(f))'$ becomes:

$$\forall x^0 \neg\neg\exists y^0 \forall z^0 G_f xyz =_0 0.$$

Applying $(\Pi^0_1\text{-}\mathsf{AC}(f))'$ is actually a single use of a primitive rule, namely the modus ponens instance

$$\frac{\varphi \quad \varphi \to \psi}{\psi},$$

since setting φ to $(\Sigma^0_1\text{-}\mathsf{LEM}(f))'$ allows us to set $\varphi \to \psi$ to $(\Pi^0_1\text{-}\mathsf{AC}(G_f))'$, what leads us to the conclusion ψ:

$$\neg\neg\exists g^1 \forall x^0, z^0 G_f x(gx)z =_0 0,$$

what is the same, unwinding G_f, as $(\Pi^0_1\text{-}\widehat{\mathsf{CA}}(f))'$.
To find the functional interpretation of $(\Pi^0_1\text{-}\widehat{\mathsf{CA}}(f))'$ first consider the functional interpretation of $(\Pi^0_1\text{-}\mathsf{AC}(G_f))'$. We can even use a partial interpretation (producing a weaker statement), since the assumption φ in the modus ponens is proved directly (satisfying the weaker statement). We obtain following representation (for f free):

$$\forall A_f^{2(0)}, T, W \Big(\forall x^0, B^1\ G_f(x, A_f xB, B(A_f xB)) =_0 0 \to$$
$$G_f(Tt_U, t_U(Tt_U), Wt_U) =_0 0 \Big).$$

Remark 1.53. This approach would be very unwise if the functional interpretation of the implication was used to compute some further results within a larger proof. Using this approach in such a case results in loss of information about the logical relation between the assumption and conclusion.

From the interpretation of Σ_1^0-LEM$'$ we know that (for f free and knowing that any term can depend on T and W)

$$\forall x, B \; G_f(x, A_f xB, B(A_f xB)) =_0 0)$$

is satisfied by setting (for f free)

$$A_f xB := \begin{cases} 1 & \text{if } f(x, B1) \neq_0 0 \\ B1 & \text{else} \end{cases}.$$

Obviously, the functional interpretation of $\forall f\ (\Pi_1^0\text{-}\widehat{\mathsf{CA}}(f))'$ just swaps the $\forall f$ and the exists quantification of the functional interpretation of $(\Pi_1^0\text{-}\widehat{\mathsf{CA}}(f))'$ adding f as a parameter to every existential variable (see also section 1.12). In particular A becomes dependent on f. We can actually define t_A, the corresponding primitive recursive term to $A(f, x, B)$ as:

$$t_A := \lambda f.\lambda x.\lambda B. \begin{cases} 1 & \text{if } f(x, B1) \neq_0 0 \\ B1 & \text{else} \end{cases},$$

what satisfies

$$t_A f = A_f.$$

It is important to mention the dependency of A on f, since this is the main difference to the ND-interpretation of AC, where the realizing terms were completely independent on this function. This means, $(\Pi_1^0\text{-}\widehat{\mathsf{CA}}(f))'$ is interpreted (keeping in mind that, as before, t_U always has T and W as parameters, and now, via A_f also f) by:

$$\forall T, W \; \Big(G_f(Tt_U, t_U(Tt_U), Wt_U) =_0 0\Big),$$

which is

$$\forall T^2, W^2 \; \Big(f(Tt_U, t_U(Tt_U)) =_0 0 \vee f(Tt_U, Wt_U) \neq_0 0\Big),$$

where $t_U :=_1 x_0$ is the same functional as in the interpretation of DNS in section 1.13, in which A is defined as above and T and W correspond to the T and W above. We summarize:

Proposition 1.54 (*ND-interpretation of* Π_1^0-$\widehat{\mathsf{CA}}(f)$). *The schema of arithmetical comprehension over numbers for purely universal formulas (for a given function* $f^{1(0)}$*)*

$$\exists g^1 \forall x^0, z^0 \; (fx(gx) =_0 0 \vee fxz \neq_0 0),$$

is ND-interpreted as follows:

$$\forall X^2, Z^2 \; \Big(f(Xt_g, t_g(Xt_g)) =_0 0 \vee f(Xt_g, Zt_g) \neq_0 0\Big),$$

where

$$t_g :=_1 \Phi_0 X u_{Z,f}^{1(1(0))} 0^0 0^1,$$

$$u_{Z,f} n^0 v^{1(0)} :=_0 \begin{cases} 1 & \textit{if } f(n, Z(v1^0)) \neq_0 0 \\ Z(v1^0) & \textit{else} \end{cases}.$$

Remark 1.55. Note that in contrast to the functional interpretation of Π^0_1-AC, where all realizing terms were independent on the function f, here the bar recursive term t_g realizing the function g has to be defined using f.

Using the ND-interpretation of Π^0_1-$\widehat{\mathsf{CA}}(f)$ we obtain also the ND-interpretations of Π^0_1-CA and Σ^0_1-CA:

Corollary 1.56 (*ND-interpretation of* Π^0_1-CA). *The schema of arithmetical comprehension over numbers for purely universal formulas*

$$\forall f^{1(0)} \exists h^1 \leq_1 \mathbf{1}^1 \forall x^0 (hx =_0 0 \leftrightarrow \forall z\, fxz \neq_0 0),$$

is ND-interpreted as follows:

$$\begin{aligned}\forall X^{0(10)(1)}, Z^{0(10)(1)}\ \big(\big(\ & t_h XZ(X(t_h XZ)(t_z XZ)) =_0 0 \to \\ & \qquad f(X(t_h XZ)(t_z XZ), Z(t_h XZ)(t_z XZ)) \neq_0 0) \ \big) \wedge \\ \big(\ & t_h XZ(X(t_h XZ)(t_z XZ)) =_0 0 \leftarrow \\ & \qquad f(X(t_h XZ)(t_z XZ), t_z XZ(X(t_h XZ)(t_z XZ))(Z(t_h XZ)(t_z XZ))) \neq_0 0\ \big)\ \big).\end{aligned}$$

The witnessing terms are

$$\begin{aligned} t_z &:= \lambda X^{0(10)(1)}, Z^{0(10)(1)}, a^0, b^0\ .\ t_g(t_f X)(t_f Z)a, \\ t_h &:= \lambda X^{0(10)(1)}, Z^{0(10)(1)}, n^0\ .\ \overleftrightarrow{f(n, t_g(t_f X)(t_f Z)n)}, \end{aligned}$$

where

$$\overleftrightarrow{n^0} :=_0 \begin{cases} 1 & \textit{if} \quad n =_0 0 \\ 0 & \textit{else} \end{cases},$$

$$t_f := \lambda X^{0(10)(1)}, g^1\ .\ X(\lambda n^0.\overleftrightarrow{f(n, gn)})(\lambda a^0, b^0.ga).$$

The remaining terms are defined as above in proposition 1.54, just we give the two type 2 arguments of t_g explicitly, i.e., t_g stands only for the term t_g of type level 3 and not for the type 1 term $t_g XZ$ as above.

Remark 1.57. We use a slightly different form of Π^0_1-CA than in definition 1.50 above in two aspects:

1. Since we can give a realizer for h^1 bounded by $\mathbf{1}^1$ we formulate the sentence directly in the stronger version with $h^1 \leq_1 \mathbf{1}^1$. We do so also in corollary 1.58.

2. In contrast to the common formulations we want h to comprehend $\forall z\, fxz \neq 0$ instead of $\forall z\, fxz = 0$. Both formulations are equivalent, though the one used in corollary 1.56 spares us some technical overhead and so improves the readability of the proof below.

Moreover, we don't strictly follow the rules of the ND-interpretation and choose the "$\forall\exists$" prenexiation over "$\exists\forall$" to start with. Again, this leads to an equivalent statement resulting in simpler and more readable terms of lower type.

Proof of corollary 1.56. Rewriting proposition 1.54 using the formally correct notation of corollary 1.56 we have that:

$$\forall \mathrm{X}^2, \mathrm{Z}^2 \ \left(f(X(t_g \mathrm{XZ}), t_g \mathrm{XZ}(X(t_g \mathrm{XZ}))) =_0 0 \lor f(X(t_g \mathrm{XZ}), Z(t_g \mathrm{XZ})) \neq_0 0 \ \right). \tag{+}$$

Given any $\mathrm{X}_0^{0(10)(1)}$, $\mathrm{Z}_0^{0(10)(1)}$ set $\mathrm{X}^2 := t_f \mathrm{X}_0$ and $\mathrm{Z}^2 := t_f \mathrm{Z}_0$ to obtain:

$$\begin{aligned} X(t_g \mathrm{XZ}) &= t_f \mathrm{X}_0(t_g \mathrm{XZ}) \\ &= \mathrm{X}_0(\lambda n.\overleftrightarrow{f(n, t_g \mathrm{XZ} n)})(\lambda a, b.t_g \mathrm{XZ} a) \\ &= \mathrm{X}_0(\lambda n.\overleftrightarrow{f(n, t_g(t_f \mathrm{X}_0)(t_f \mathrm{Z}_0) n)})(\lambda a, b.t_g(t_f \mathrm{X}_0)(t_f \mathrm{Z}_0) a) \\ &= \mathrm{X}_0(t_h \mathrm{X}_0 \mathrm{Z}_0)(t_z \mathrm{X}_0 \mathrm{Z}_0) \end{aligned} \tag{*}$$

and analogously $Z(t_g \mathrm{XZ}) = \mathrm{Z}_0(t_h \mathrm{X}_0 \mathrm{Z}_0)(t_z \mathrm{X}_0 \mathrm{Z}_0)$.

- Suppose we have $t_h \mathrm{X}_0 \mathrm{Z}_0(\mathrm{X}_0(t_h \mathrm{X}_0 \mathrm{Z}_0)(t_z \mathrm{X}_0 \mathrm{Z}_0)) =_0 0$.
  It follows by (*) that $t_h \mathrm{X}_0 \mathrm{Z}_0(X(t_g \mathrm{XZ})) =_0 0$ and by definition of t_h that

  $$f(X(t_g \mathrm{XZ}), t_g(t_f \mathrm{X}_0)(t_f \mathrm{Z}_0)(X(t_g \mathrm{XZ}))) \neq 0.$$

  By definition of X and Z we get $f(X(t_g \mathrm{XZ}), t_g \mathrm{XZ}(X(t_g \mathrm{XZ}))) \neq 0$ which implies

  $$f(X(t_g \mathrm{XZ}), Z(t_g \mathrm{XZ})) \neq 0$$

  by (+) and

  $$f(\mathrm{X}_0(t_h \mathrm{X}_0 \mathrm{Z}_0)(t_z \mathrm{X}_0 \mathrm{Z}_0), \mathrm{Z}_0(t_h \mathrm{X}_0 \mathrm{Z}_0)(t_z \mathrm{X}_0 \mathrm{Z}_0)) \neq 0$$

  by (*).

- On the other hand, let

  $$f(\mathrm{X}_0(t_h \mathrm{X}_0 \mathrm{Z}_0)(t_z \mathrm{X}_0 \mathrm{Z}_0), t_z \mathrm{X}_0 \mathrm{Z}_0(\mathrm{X}_0(t_h \mathrm{X}_0 \mathrm{Z}_0)(t_z \mathrm{X}_0 \mathrm{Z}_0))(\mathrm{Z}_0(t_h \mathrm{X}_0 \mathrm{Z}_0)(t_z \mathrm{X}_0 \mathrm{Z}_0))) \neq 0.$$

  Unwinding t_z we get

  $$f(\mathrm{X}_0(t_h \mathrm{X}_0 \mathrm{Z}_0)(t_z \mathrm{X}_0 \mathrm{Z}_0), t_g(t_f \mathrm{X}_0)(t_f \mathrm{Z}_0)(\mathrm{X}_0(t_h \mathrm{X}_0 \mathrm{Z}_0)(t_z \mathrm{X}_0 \mathrm{Z}_0))) \neq 0.$$

  Hence, we obtain $t_h \mathrm{X}_0 \mathrm{Z}_0(\mathrm{X}_0(t_h \mathrm{X}_0 \mathrm{Z}_0)(t_z \mathrm{X}_0 \mathrm{Z}_0)) = 0$.

□

The schema $(\Sigma_1^0\text{-CA})'$ is D-interpreted in a similar way.

Corollary 1.58 (*ND-interpretation of* Σ_1^0*-CA(f)). The schema of arithmetical comprehension over numbers for purely existential formulas (for a given function* $f^{1(0)}$*)*

$$\exists h^1 \leq_1 \mathbf{1}^1 \forall x^0 (hx =_0 0 \leftrightarrow \exists z\, fxz =_0 0),$$

is ND-interpreted as follows:

$$\begin{aligned} \forall \mathrm{X}^{0(10)(1)}, \mathrm{Z}^{0(10)(1)} \ (\ & t_h \mathrm{XZ}(X(t_h \mathrm{XZ})(t_z \mathrm{XZ})) =_0 0 \leftarrow \\ & \qquad f(X(t_h \mathrm{XZ})(t_z \mathrm{XZ}), Z(t_h \mathrm{XZ})(t_z \mathrm{XZ})) \neq_0 0) \ \land \\ & t_h \mathrm{XZ}(X(t_h \mathrm{XZ})(t_z \mathrm{XZ})) =_0 0 \rightarrow \\ & \qquad f(X(t_h \mathrm{XZ})(t_z \mathrm{XZ}), t_z \mathrm{XZ}(X(t_h \mathrm{XZ})(t_z \mathrm{XZ}))(Z(t_h \mathrm{XZ})(t_z \mathrm{XZ}))) \neq_0 0 \ ). \end{aligned}$$

As above in corollary 1.56, the witnessing terms are:

$$t_z := \lambda X^{0(10)(1)}, Z^{0(10)(1)}, a^0, b^0 \,.\, t_g(t_f X)(t_f Z)a,$$
$$t_h := \lambda X^{0(10)(1)}, Z^{0(10)(1)}, n^0 \,.\, \overleftrightarrow{f(n, t_g(t_f X)(t_f Z)n)}.$$

The term t_g corresponds to the term defined in proposition 1.54. The only difference is that we give the two type 2 arguments of t_g explicitly, i.e., t_g stands only for the term t_g of type level 3 and not for the type 1 term $t_g XZ$ as above. We redefine $\overleftrightarrow{n^0}$ as:

$$\overleftrightarrow{n^0} :=_0 \begin{cases} 0 & \text{if} \quad n =_0 0 \\ 1 & \text{else} \end{cases},$$

what slightly changes the meaning of t_f (and t_h) compared to corollary 1.56. However, we keep the exact syntactic form for t_f as before:

$$t_f := \lambda X^{0(10)(1)}, g^1 \,.\, X(\lambda n^0.\overleftrightarrow{f(n, gn)})(\lambda a^0, b^0.ga).$$

Proof. The proof is essentially the same as the proof of 1.56. By (+) and (*) we refer to the statements (+) and (*) in that proof. The notation and definitions of X and Z remain syntactically the same as above as well.

- Suppose we have $t_h X_0 Z_0(X_0(t_h X_0 Z_0)(t_z X_0 Z_0)) =_0 0$. It follows by (*) that

  $$t_h X_0 Z_0(X(t_g XZ)) =_0 0$$

  and by definition of t_h that

  $$f(X(t_g XZ), t_g(t_f X_0)(t_f Z_0)(X(t_g XZ))) = 0.$$

  By (*) and definition of X and Z we get

  $$f(X_0(t_h X_0 Z_0)(t_z X_0 Z_0), t_g(t_f X_0)(t_f Z_0)(X_0(t_h X_0 Z_0)(t_z X_0 Z_0))) =$$
  $$f(X_0(t_h X_0 Z_0)(t_z X_0 Z_0), t_z X_0 Z_0(X_0(t_h X_0 Z_0)(t_z X_0 Z_0))(Z_0(t_h X_0 Z_0)(t_z X_0 Z_0)))) = 0.$$

- On the other hand, let $f(X_0(t_h X_0 Z_0)(t_z X_0 Z_0), Z_0(t_h X_0 Z_0)(t_z X_0 Z_0)) = 0$. By (*) it means $f(X(t_g \underline{U}), Z(t_g \underline{U})) = 0$, what implies $f(X(t_g XZ), t_g XZ(X(t_g XZ))) = 0$ by (+). Using (*) and the definition of X, Z, and t_h we obtain

  $$f(X_0(t_h X_0 Z_0)(t_z X_0 Z_0), t_g(t_f X_0)(t_f Z_0)(X_0(t_h X_0 Z_0)(t_z X_0 Z_0))) =$$
  $$t_h X_0 Z_0(X_0(t_h X_0 Z_0)(t_z X_0 Z_0))) = 0.$$

□

Remark 1.59. We mentioned earlier that we will use M^ω and the stability of atomic formulas modulo double negation silently to obtain more readable results. In addition, in this case we use two modifications to obtain the ND-interpretation of Σ^0_1-CA. Each modification leads to an intuitionistically equivalent formula:

$$\cfrac{\cfrac{\neg\neg\exists h^1 \forall x^0(hx =_0 0 \leftrightarrow \neg\neg\exists z\, fxz =_0 0)}{\neg\neg\exists h^1 \forall x^0(hx =_0 0 \leftrightarrow \neg\forall z\, \neg(fxz =_0 0))}\ \text{Lemma 1.5}}{\neg\neg\exists h^1 \forall x^0(hx \neq_0 0 \leftrightarrow \forall z\, fxz \neq_0 0)}\ \text{Definition 1.3}.$$

The ND-interpretations of Π^0_1-CA and Σ^0_1-CA could be simplified even more. Namely, we have $\forall X^{0(10)(1)}, Z^{0(10)(1)}, a^0, b^0 t_z XZab =_0 t_z XZa0$, i.e. the term $t_z XZ$ does not depend on its second type 0 argument. Hence, we could define the terms t_z and t_f as:

$$t_z := \lambda X^{2(1)}, Z^{2(1)}, a^0 \,.\, t_g(t_f X)(t_f Z)a$$
$$t_f := \lambda X^{2(1)}, g^1 \,.\, X(\lambda n^0.\overleftrightarrow{f(n, gn)})g$$

and reformulate the corollaries 1.56 and 1.58 equivalently for all $X^{2(1)}$ and $Z^{2(1)}$ using

$$t_z XZ(X(t_h XZ)(t_z XZ))$$

instead of $t_z XZ(X(t_h XZ)(t_z XZ))(Z(t_h XZ)(t_z XZ))$.

Summary

The main result, which we will need also in later sections is the ND-interpretation of CA (Corollary 1.58), we reformulate is here using the more compact Sh-interpretation:

Theorem 1.60 (Sh-interpretation of Σ^0_1-CA). *The schema of arithmetical comprehension over numbers for purely existential formulas (for a given function $f^{1(0)}$)*

$$\exists h^1 \forall x^0 (hx =_0 0 \leftrightarrow \exists z\, fxz =_0 0),$$

is Sh-interpreted as follows (using the clause (S7) for the conjunction hidden in '$\leftrightarrow$'):*

$$\begin{aligned}\forall X^{0(10)(1)}, Z^{0(10)(1)}\ [(\ & t_h XZ(X(t_h XZ)(t_z XZ)) =_0 0 \leftarrow \\ & \quad f(X(t_h XZ)(t_z XZ), Z(t_h XZ)(t_z XZ)) =_0 0)\ \wedge \\ (& t_h XZ(X(t_h XZ)(t_z XZ)) =_0 0 \rightarrow \\ & \quad f(X(t_h XZ)(t_z XZ), t_z XZ(X(t_h XZ)(t_z XZ))(Z(t_h XZ)(t_z XZ))) =_0 0\ )].\end{aligned}$$

The witnessing terms are:

$$t_z := \lambda X^{0(10)(1)}, Z^{0(10)(1)}, a^0, b^0 \,.\, t_g(t_f X)(t_f Z)a,$$
$$t_h := \lambda X^{0(10)(1)}, Z^{0(10)(1)}, n^0 \,.\, \overleftrightarrow{f(n, t_g(t_f X)(t_f Z)n)},$$

where $t_f := \lambda X^{0(10)(1)}, g^1 \,.\, X(\lambda n^0.\overleftrightarrow{f(n, gn)})(\lambda a^0, b^0.ga)$ *and* $\overleftrightarrow{n^0} :=_0 \begin{cases} 0 & \text{if } n =_0 0 \\ 1 & \text{else} \end{cases}$.

The term t_g corresponds to the term defined in proposition 1.54. The only difference is that we give the two type 2 arguments of t_g explicitly, i.e., t_g stands only for the term t_g of type level 3 and not for the type 1 term $t_g XZ$ as above.

1.16 Weak König's Lemma

The König's lemma as such is mostly known in the following form:

Every infinite, though finitary branching, tree has an infinite path.

However, this version is fairly strong and it can be shown that it is equivalent to CA^0 if we allow arbitrary formulas for the decision of belonging to the tree (see e.g. [126]). If the labels for branches are bounded by a function α depending only on the current node *and* the tree is defined by a quantifier-free (or purely universal) formula we get a significantly weaker tool, which is intuitionistically equivalent to WKL, where we restrict the lemma to binary trees only (i.e. $\alpha(\cdot) \leq 1$). Even so, namely for quantifier-free binary trees, the lemma becomes equivalent to CA^0_{ar} if we ask for the left-most or some other concrete infinite path instead of just some infinite path. This means we need all three weakenings:

- The decision criteria, $\varphi(x^0)$, for an initial segment of a sequence of natural numbers, x, to belong to the tree defined by the characteristic function f must be a Π^0_1 formula (we can allow one for-all-quantifier since it has no essential influence on the structure of the lemma - see definition 1.61 and proposition 1.63). If we allow arbitrary formulas, we get even full CA^0 (regardless of any bounds on the number of branches greater than 1).
- For each node, the labels of its branches must be bounded by a well defined type one function depending only on the height of the node. Otherwise, we get Π^0_1-CA and by iteration CA^0_{ar} even for quantifier-free decision criteria.
- There can't be any additional (infinite) demands on the infinite path except for its existence.

We consider the following definition:

Definition 1.61 (WKL(φ))**.** For a given φ we define the following theorem. Every infinite binary tree given by the decision criteria φ has an infinite path,

$$\mathsf{WKL}(\varphi) \quad : \quad \mathsf{BinTree}(\varphi) \wedge \forall k \mathsf{Unbounded}(\varphi, k) \rightarrow \exists b \Big(\mathsf{BinFunc}(b) \wedge \forall k\ \varphi(\bar{b}(k)) \Big),$$

where

$$\begin{aligned} \mathsf{BinFunc}(b) &:\equiv \quad \forall n^0 (b(n) =_0 0 \vee b(n) =_0 1), \\ \mathsf{BinTree}(\varphi) &:\equiv \quad \forall s(\ \varphi(s) \rightarrow\ s \in \{0,1\}^{<\omega} \wedge \forall t \subseteq s\ \varphi(t)\ ), \\ \mathsf{Unbounded}(\varphi, k^0) &:\equiv \quad \exists s \in \{0,1\}^k\ \varphi(s). \end{aligned}$$

Furthermore, we define the schema Π^0_n-WKL, as the union of WKL(φ), where φ is a Π^0_n formula. Also, we write Π^0_n-WKL(φ) to indicate that we mean the concrete instance WKL(φ) and that φ is a Π^0_n formula.

Note that, for every fixed $n \in \mathbb{N}$, we can always reformulate the schema Π^0_n-WKL as a single 2nd-order axiom. We will use this fact implicitly. However, in the special case for quantifier-free φ we define explicitly:

Definition 1.62 (WKL $\equiv \forall f$WKL(f) see also [126])**.** Every infinite binary tree, given by the characteristic function f, has an infinite path:

$$\mathsf{WKL}(f) \quad : \quad \mathsf{BinTree}_K(f) \wedge \forall k \exists x (\,\mathrm{lh}(x) =_0 k \wedge f(x) =_0 0) \rightarrow$$
$$\exists b \leq_1 \mathbf{1}^1 \Big(\forall k\, f(\bar{b}(k)) =_0 0 \Big),$$

where

$$\mathsf{BinTree}_K(f^1) :\equiv \forall x, y(f(x * y) =_0 0 \to fx =_0 0) \ \wedge\ \forall x, n(f(x * \langle n \rangle) =_0 0 \to n \leq_0 1).$$

We mentioned earlier that the schema Π^0_1-WKL is essentially equivalent to WKL. this is proved by the proposition 1.63 below, using the fact that in all systems used in this thesis there is a suitable f^1 for any quantifier-free $\varphi_{\mathsf{QF}}(\underline{n})$ with only the type 0 parameters $\underline{n}$ free s.t. $f(\underline{n}) =_0 0 \Leftrightarrow \varphi_{\mathsf{QF}}(\underline{n})$ (and vice-versa).

Proposition 1.63.

$$\mathsf{WE\text{-}HA}^\omega \vdash \Pi^0_1\text{-}\mathsf{WKL} \leftrightarrow \mathsf{WKL}$$

Proof (see also [114]).

- Π^0_1-WKL $\to$ WKL:
  If we have WKL(φ) for Π^0_1 formulas φ, it surely holds for quantifier-free formulas especially for $\varphi :\equiv f(n^0) =_0 0$.
- Π^0_1-WKL $\leftarrow$ WKL:
  Since $\varphi \in \Pi^0_1$, $\varphi(s)$ can be written as $\forall n^0 \varphi_0(s, f, n)$. Next we involve two tricks in definition of a quantifier-free version of φ:

  $$\tilde{\varphi}(s, f) :\equiv \forall n^0 <_0 \mathrm{lh}(s), t \subseteq s(\varphi_0(f, t, n) \wedge s(n) \leq_0 1).$$

  1. Since we get $\forall n^0 <_0 \mathrm{lh}(s)\varphi_0(f, t, n)$ for-all s in the conclusion of WKL, the number n is in fact unbounded.
  2. We guarantee BinTree$(\tilde{\varphi})$ as a free property for any $\tilde{\varphi}$. The tree property by forcing

     $$\forall t \subseteq s\ \varphi_0(f, t, n)$$

     and the binary property by

     $$\forall n_0 <_0 \mathrm{lh}(s), t \subseteq s\ s(n) \leq_0 1.$$

  Now, suppose

  $$\mathsf{BinTree}(\varphi) \wedge \forall k \mathsf{Unbounded}(\varphi, k)$$

  i.e. we have (by $\forall k$Unbounded(φ, k)):

  $$\forall k\ \exists s \in \{0,1\}^k\ \forall n\ \ \varphi_0(f, n, s). \qquad (+)$$

  By BinTree(φ) and (+) we get:

  $$\forall k \exists s \in \{0,1\}^k (\forall n^0 <_0 \mathrm{lh}(s), t \subseteq s(\varphi_0(f, t, n) \wedge s(n) \leq_0 1)),$$

  i.e. we have

  $$\forall k \exists s \in \{0,1\}^k \tilde{\varphi}(f, s),$$

  and classically also

  $$\forall k \mathsf{Unbounded}(\tilde{\varphi}, k).$$

  By the steps above we showed so far (using trick (2) to get BinTree$(\tilde{\varphi})$):

  $$\mathsf{BinTree}(\varphi) \wedge \forall k \mathsf{Unbounded}(\varphi, k) \quad \to \quad \mathsf{BinTree}(\tilde{\varphi}) \wedge \forall k \mathsf{Unbounded}(\tilde{\varphi}, k).$$

We can define in WE-HA$^\omega$ the primitive recursive term for the following type one function g^1:

$$g(s) :=_0 \begin{cases} 0^0 & \text{if} \tilde{\varphi}(f,s) \\ 1^0 & \text{else} \end{cases}.$$

To get equivalently:

$$\mathsf{BinTree}(f) \wedge \forall k \mathsf{Unbounded}(f,k) \quad \rightarrow \quad \mathsf{BinTree}_K(g) \wedge \forall k \mathsf{Unbounded}(g,k).$$

Applying WKL(g) to the conclusion we get:

$$\exists b \leq_1 \mathbf{1}^1 \forall k\, g(\bar{b}(k)) =_0 0$$

what is, by definition of g, the same as

$$\exists b \leq_1 \mathbf{1}^1 \forall k\, \tilde{\varphi}(\bar{b}(k), f),$$

what is, by definition of $\tilde{\varphi}$, the same as

$$\exists b^1 \leq_1 \mathbf{1}^1 \forall k^0\, \forall n^0 <_0 k \forall t \subseteq \bar{b}(k)\;\; \varphi_0(f,t,n).$$

This implies (using trick (1) from the definition of $\tilde{\varphi}$):

$$\exists b^1 \leq_1 \mathbf{1}^1 \forall k^0 \forall n^0 \forall t \subseteq \bar{b}(k) \varphi_0(f,t,n).$$

Observing that $\forall k \forall t \subseteq \bar{b}(k)\; \psi(t)$ is equivalent to $\forall k \psi(\bar{b}(k))$ for arbitrary formulas ψ we finally obtain:

$$\exists b^1 \leq_1 \mathbf{1}^1 \forall k^0\, (\forall n^0\, \varphi_0(f,\bar{b}(k),n)) \quad \equiv \quad \exists b^1 \leq_1 \mathbf{1}^1 \forall k^0\, \varphi(f,\bar{b}(k)).$$

□

Remark 1.64. Where the necessity of trick (1) in the definition of $\tilde{\varphi}$ is easy to see, one might think that the trick (2) is not needed. This is not true. The following example shows that in general we even *do not* have $\mathsf{BinTree}(\forall n \varphi_0(f,s,n)) \rightarrow \mathsf{BinTree}(\forall n^0 <_0 \mathrm{lh}(s)(\varphi_0(f,s,n) \wedge s(n) \leq_0 1)$. For some given constant number $c^0 \in \mathbb{N}$, define

$$\varphi_0(f,s,n) :\equiv \quad s \in \{0,1\}^{\mathrm{lh}(s)} \wedge (s(0) =_0 0 \vee s(0) > c + 1 + n - 2 * \mathrm{lh}(s)).$$

While $\forall n \varphi_0(f,s,n)$ defines the tree consisting exactly of all binary sequences starting with 0, the formula $\forall n^0 <_0 \mathrm{lh}(s) \varphi_0(f,s,n)$ is true for all such sequences but also for all binary sequences of length greater than c. In other words, any sequence of length e.g. $c+1$ starting with 1 is a counterexample for $\mathsf{BinTree}(\forall n^0 <_0 \mathrm{lh}(s)(\varphi_0(f,s,n) \wedge s(n) \leq_0 1)$.

Kohlenbach showed in his work (see [61], [71]), that WKL is intuitionistically equivalent to a sentence in Δ as defined in theorem 1.29. We sketch this result in 1.16. This equivalence actually suggests that the lemma will not have any significant effect on the final complexity of the realizers of sentences whose proofs are based on WKL in common systems.
There are several ways to prove this fact. Howard gives a technical argument using restricted Bar Recursion in [53] which we follow in 1.16.

Kohlenbach's WKL′ as a Sentence Δ

First, observe that WKL is equivalent (in $\widehat{\text{WE-HA}^{\omega}}\!\upharpoonright$) to:

$$\mathsf{WKL}_{K1} \quad : \quad \forall f, g \left(\mathsf{BinTree}_K(f) \wedge \forall k\big(\,\mathrm{lh}(\bar{g}k) =_0 k \wedge f(\bar{g}k) =_0 0\big) \to \exists b \leq_1 \mathbf{1}^1 \forall k^0 \big(f(\bar{b}k) =_0 0\big)\right).$$

Next, we define for any type 1 function f^1 the constructions $\hat{f}$ and f_g:

Definition 1.65.

$$\hat{f}n := \begin{cases} fn & \text{if} fn \neq 0 \ \vee \ \big(\forall k, l(k * l = n \to fk = 0) \wedge \forall i < \mathrm{lh}(n)\ (n_i \leq 1)\big) \\ 1^0 & \text{else} \end{cases}$$

$$f_g n := \begin{cases} fn & \text{if } f\big(g(\mathrm{lh}(n))\big) = 0 \wedge \mathrm{lh}\left(g(\mathrm{lh}(n))\right) = \mathrm{lh}(n) \\ 0^0 & \text{else} \end{cases}.$$

Remark 1.66. We defined already a construction $\hat{o}$ for a type 1 object in 1.14. It should be obvious which meaning should be assigned from the context. From now on, we will use it mostly in the sense of 1.65.

Now, we are able to define

Definition 1.67.

$$\mathsf{WKL}' \quad : \quad \forall f^1, g^1 \exists b \leq_1 \mathbf{1}^1 \forall k^0 \left(\widehat{(f)}_g(\bar{b}k) =_0 0\right).$$

To obtain:

Proposition 1.68. *The sentence* WKL′ *is in* Δ *as defined in theorem 1.29 and:*

$$\mathsf{WE\text{-}HA}^{\omega} \vdash \mathsf{WKL} \leftrightarrow \mathsf{WKL}'.$$

The complete proof of an even stronger result, namely of the equivalence under $\widehat{\text{WE-HA}^{\omega}}\!\upharpoonright$, can be found in [71]. In that book, Kohlenbach also proves the following lemma:

Lemma 1.69.

$$\widehat{\text{WE-HA}^{\omega}}\!\upharpoonright \ \vdash \ \forall f\, \mathsf{BinTree}_K(\hat{f}) \tag{1}$$

$$\widehat{\text{WE-HA}^{\omega}}\!\upharpoonright \ \vdash \ \forall f\, \left(\mathsf{BinTree}_K(f) \to f =_1 \hat{f}\right) \tag{2}$$

$$\widehat{\text{WE-HA}^{\omega}}\!\upharpoonright \ \vdash \ \forall f, g \forall n \exists x\, \left(\,\mathrm{lh}(x) = n \wedge f_g(x) = 0\right) \tag{3}$$

$$\widehat{\text{WE-HA}^{\omega}}\!\upharpoonright \ \vdash \ \forall f, g \left(\forall n\big(\,\mathrm{lh}(gn) = n \wedge f(gn) = 0\big) \to f_g =_1 f\right) \tag{4}$$

For the purpose of this thesis mainly WKL ← WKL′ is interesting:
Proof. Assume $\mathsf{BinTree}_K(f)$ and $\forall k \exists n(\mathrm{lh}(n) = k \wedge fn = 0))$. Then

$$\forall k \exists n \leq \overline{\mathbf{1}^1} k(\mathrm{lh}(n) = k \wedge fn = 0)).$$

Define primitive recursive in f:

$$gk := \begin{cases} \min n \leq \overline{1^1}k \ (\mathrm{lh}(n) = k \wedge fn = 0) & \text{if such an } n \text{ exists} \\ 0^0 & \text{else} \end{cases},$$

to obtain $\forall k(\mathrm{lh}(gk) = k \wedge f(gk) = 0)$.
Using lemma 1.69 and other Kohlenbach's results from [71] we get $f_g =_1 f$ and $f =_1 \hat{f}$.
This proves $\widehat{(\hat{f})_g} =_1 f$.
Finally WKL' yields $\exists b \leq_1 \mathbf{1}^1 \forall k^0 (f(\bar{b}k) = 0)$.

□

Using the fixed construction for g from the proof above, we can define a fixed term s of $\mathsf{WE\text{-}HA}^\omega$, and formulate yet another version of WKL:

Definition 1.70. $\mathsf{WKL}_s \equiv \forall f \mathsf{WKL}_s(f)$, where

$$\mathsf{WKL}_s(f) \quad :\equiv \quad \exists b \leq_1 \mathbf{1}^1 \forall k^0 \left(\widehat{(\hat{f})_s}(\bar{b}k) =_0 0 \right).$$

By the proof of proposition 1.68, we still have:

Proposition 1.71.

$$\mathsf{WE\text{-}HA}^\omega \vdash \mathsf{WKL}_s \rightarrow \mathsf{WKL}.$$

Howard's D-interpretation of WKL

We cannot D-interpret the lemma directly. We could apply the negative translation to the lemma and then go for the D-interpretation, but it turns out that the proof is simpler when we go for the ND-interpretation of a classically equivalent formula, the so called FAN principle. In this section, let f^1 be a characteristic function of a tree, x^0 and y^0 encodings of finite $\{0,1\}$-sequences, and b^1 a $\{0,1\}$-sequence (i.e. $b \in \{0,1\}^\omega$). We define the FAN principle as follows (note that the hat construction does not affect the general meaning - see proposition 1.68 - however, it does significantly simplify the terms as it eliminates some quantifiers):

Definition 1.72 ($\mathsf{FAN} \equiv \forall f \mathsf{FAN}(f)$)**.** We define the FAN principle for a given function f as

$$\mathsf{FAN}(f) \quad :\equiv \quad \forall b \exists j \, \mathsf{Sec}(\bar{b}j) \ \rightarrow \ \exists k \forall x \left(\mathrm{lh}(x) \geq k \rightarrow \mathsf{Sec}(x) \right),$$

where Sec means *secured*:

$$\mathsf{Sec}(x) \quad :\equiv \quad \hat{f}(x) \neq 0.$$

In [53], Howard showed that the D-interpretation of WKL (using the classically equivalent FAN) can be obtained using only the so called restricted Bar Recursion (defined in the same publication, see also definition 1.42). We will present this proof filling in some details (the proofs of lemma 1.74 and lemma 1.75 as well as the witness for the binary sequence b).
The secured property is conservative over extensions, i.e. for any two finite binary sequences x and y we have:

$$x \subseteq y \rightarrow \left(\mathsf{Sec}(x) \rightarrow \mathsf{Sec}(y) \right). \qquad \text{(SC)}$$

Hence, the conclusion of $\mathsf{FAN}(f)$ is equivalent to a Σ_1^0 formulation (as the for-all quantifier can be bounded):

$$\forall b \exists j\, \mathsf{Sec}(\bar{b}j) \;\rightarrow\; \exists k \underbrace{\forall x \in \{0,1\}^k\, \mathsf{Sec}(x)}_{\text{quantifier-free}}.$$

Moreover, in the presence of the Markov principle, the negative interpretation does not change the formula. Already in intuitionistic logic, $\neg\neg(A \rightarrow B)$ is equivalent to $A \rightarrow \neg\neg B$. So, the negative interpretation becomes

$$\forall b\, \neg\neg\, \exists j\, \mathsf{Sec}(\bar{b}j) \;\rightarrow\; \neg\neg(\exists k \forall\, x \in \{0,1\}^k\, \mathsf{Sec}(x)).$$

By the Markov principle we get

$$\forall b\, \exists j\, \neg\neg \mathsf{Sec}(\bar{b}j) \;\rightarrow\; \exists k\, \neg\neg \forall\, x \in \{0,1\}^k\, \mathsf{Sec}(x)$$

and because of the stability of quantifier-free formulas modulo double negation, we get just the very same sentence we started with. In summary we have:

$$\begin{aligned}
\mathsf{WE\text{-}PA}^\omega &\vdash\; \mathsf{WKL} \leftrightarrow \mathsf{FAN},\\
\mathsf{WE\text{-}HA}^\omega &\vdash\; (\mathsf{WKL})' \leftrightarrow (\mathsf{FAN})',\\
\mathsf{WE\text{-}HA}^\omega + \mathsf{M}^\omega &\vdash\; \mathsf{FAN} \leftrightarrow (\mathsf{FAN})',\\
\mathsf{WE\text{-}HA}^\omega + \mathsf{M}^\omega &\vdash\; (\mathsf{WKL})' \leftrightarrow \mathsf{FAN},
\end{aligned}$$

and therefore also:

$$\mathsf{WE\text{-}HA}^\omega + \mathsf{M}^\omega \vdash\; (\mathsf{WKL})^{ND} \leftrightarrow (\mathsf{FAN})^D.$$

So FAN is precisely the form of WKL we want to D-interpret.
By quantifier-free Axiom of Choice $\mathsf{QF\text{-}AC}^{1,0}$ we get from $\forall b \exists j\, \mathsf{Sec}(\bar{b}j)$:

$$\exists A \forall b\, \mathsf{Sec}(\bar{b}(Ab)).$$

Now, define K_A as follows:

$$K_A x := \begin{cases} 0 & \text{if } A[x] < \mathrm{lh}(x) \\ 1 + \max\{K_A(x * 0), K_A(x * 1)\} & \text{else} \end{cases}.$$

Further define:

$$\begin{aligned}
&\mathsf{BarSec}(k,x) :\equiv \forall y\, \Big((x \subseteq y \wedge \mathrm{lh}(x) + k = \mathrm{lh}(y)) \;\rightarrow\; \mathsf{Sec}(y)\Big),\\
&\mathsf{BarSec}_A(x) :\equiv \mathsf{BarSec}(K_A x, x).
\end{aligned}$$

Remark 1.73. $\mathsf{BarSec}(k,x)$ is the predicate for: Every finite extension of x with the length $\mathrm{lh}(x) + k$ is secured.

We will make use of the following two lemmas (in both cases we consider some functional A for which $\forall b\, \mathsf{Sec}(\bar{b}(Ab))$ holds, moreover, w.l.o.g. it may be assumed that A is computable since it was obtained by $\mathsf{QF\text{-}AC}^{1,0}$):

Lemma 1.74.

$$A[x] < \mathrm{lh}(x) \to \mathsf{BarSec}_A(x)$$

Proof. We assume:

$$A[x] < \mathrm{lh}(x). \quad (*)$$

By definition of K_A we get:

$$K_A x = 0. \quad (**)$$

From $\forall b\, \mathsf{Sec}(\bar{b}(Ab))$ we get

$$\mathsf{Sec}(\overline{[x]}(A[x])).$$

By (*) we know that

$$\overline{[x]}(A[x]) \subseteq x$$

hence by (SC) it follows

$$\mathsf{Sec}(x).$$

Using (**) and the definitions of BarSec and BarSec_A we get:

$$\mathsf{Sec}(x) \equiv \mathsf{BarSec}(0, x) \equiv \mathsf{BarSec}_A(x).$$

□

Lemma 1.75.

$$\mathsf{BarSec}_A(x * 0) \wedge \mathsf{BarSec}_A(x * 1) \to \mathsf{BarSec}_A(x)$$

Proof. Assume:

$$\mathsf{BarSec}_A(x * 0) \wedge \mathsf{BarSec}_A(x * 1). \quad (*)$$

If $A[x] < \mathrm{lh}(x)$, then by lemma 1.74 we have $\mathsf{BarSec}_A(x)$. So, using the definition of K_A we may assume w.l.o.g:

$$K_A x = 1 + max\{K_A(x * 0), K_A(x * 1)\}.$$

Obviously we have:

$$K_A x > K_A(x * 0) \ \wedge \ K_A x > K_A(x * 1).$$

So by (*) we know that for some number $m \leq \mathrm{lh}(x) + K_A(x)$ all extensions of $(x * 0)$ and $(x * 1)$ of the fixed length m are secured. Therefore all extensions of x with fixed length m are secured. However, this is expressed as $\mathsf{BarSec}(K_A x, x) \equiv \mathsf{BarSec}_A(x)$.

□

Since $\mathsf{BarSec}_A(x)$ can be written as a quantifier-free formula for any given x, and hence is primitive recursive in x, we can use primitive recursion on the contrapositive of lemma 1.75:

$$\neg\mathsf{BarSec}_A(x) \ \to \ \neg\big(\mathsf{BarSec}_A(x * 0) \wedge \mathsf{BarSec}_A(x * 1)\big),$$

to obtain a function $g \leq \mathbf{1}$, s.t.

$$\neg\mathsf{BarSec}_A(\varnothing) \to \forall j \neg\mathsf{BarSec}_A(\bar{g}j).$$

However, we can prove the existence of a j s.t. $\mathsf{BarSec}_A(\bar{g}j)$, using only restricted Bar Recursion, and thereby proving $\mathsf{BarSec}_A(\varnothing)$. To do so, we follow another publication from Howard, namely [50], where he proves an, by lemma 1.74, even stronger lemma (for the same kind of functional A as above):

Lemma 1.76. *By restricted bar recursion plus primitive recursion, we can define θ_A, s.t. for all b it holds:*

$$\exists k \leq \theta_A b\varnothing \quad A[\bar{b}k] < k.$$

Remark 1.77. In fact the functional θ_A is given simply by $\theta_A b\varnothing := \mu k.A[\bar{b}k] \leq k$. Since we have to work in a model justifying bar recursion and, as discussed e.g. by Kohlenbach in [71], we can define the μ-operator via bar recursion this functional is well defined in all such models, e.g. $\mathcal{C}^\omega$ or $\mathcal{M}^\omega$. Howard's proof of lemma 1.76 is analyzed in the first appendix of [106]. It is rather technical and not essential at this point.

Lemma 1.76 especially implies $\forall g^1 \leq \mathbf{1}\, \exists k^0\ A[\bar{g}k] < k$, which implies $\forall g^1 \leq \mathbf{1}\, \exists k^0$ $\mathsf{BarSec}_A(\bar{g}k)$ by lemma 1.74. So, as mentioned above, using this result we obtain:

$$\mathsf{BarSec}_A(\varnothing) \leftrightarrow \mathsf{BarSec}(K_A\varnothing, \varnothing).$$

This is, by definition, just another form of:

$$\forall x \left((\varnothing \subseteq x \wedge K_A(\varnothing) = \mathrm{lh}(x)) \rightarrow \mathsf{Sec}(x) \right).$$

So, finally we obtain:

$$\forall x \left(\mathrm{lh}(x) \geq K_A(\varnothing) \rightarrow \mathsf{Sec}(x) \right).$$

What completes the proof that $K_A(\varnothing)$ is the ND-realizer of k in $\mathsf{FAN}(f)$.

To complete the solution of the D-interpretation we still have to find the functional B which satisfies:

$$\forall A, x \left(\mathsf{Sec}(\overline{B}(AB)) \rightarrow (\mathrm{lh}(x) \geq K_A(\varnothing) \rightarrow \mathsf{Sec}(x)) \right).$$

Fortunately, this is easily done. Let y^0 encode either a binary sequence of length at most $K_A(\varnothing)$ which is in the tree and satisfies $A[y] < \mathrm{lh}(y)$ or, if such a sequence doesn't exist, the empty sequence. Now we can define the functional B as the infinite extension of this sequence, $[y]$. See also the theorem 1.81 below.
The same functionals interpret the standard (positive) formulation of WKL, which is just the contrapositive of the outer implication:

$$\forall A, x\ \left((\mathrm{lh}(x) \geq K_A(\varnothing) \wedge x \in f) \rightarrow f(\overline{B}(AB)) = 0 \right).$$

We summarize the above formally in the following theorem.

Theorem 1.78 (The ND-interpretation of $\mathsf{WKL}(f)$). *The Weak König's lemma for binary trees given by an arithmetic characteristic function f*

$$\mathsf{BinTree}(f) \wedge \forall k \mathsf{Unbounded}(f,k) \to \exists b \Big(\mathsf{BinFunc}(b) \wedge \forall k\, f(\bar{b}(k)) = 0 \Big)$$

is, provably in $\mathsf{WE\text{-}HA}^\omega + \mathsf{QF\text{-}AC}^{1,0} + \mathsf{B'}_1$, *ND-interpreted as follows:*

$$\forall A, x \ \Big(\ (\mathrm{lh}(x) \geq K_A(\varnothing) \wedge x \in f) \to \ f(\overline{B}(AB)) = 0 \ \Big),$$

where

$$B := B(A) := \big[F_f(K_A(\varnothing))\big],$$

$$F_f(n) := F_f(A,n) := \begin{cases} \min_0\{b^0 \ : \ \mathrm{lh}(b^0) \leq n, f(b) =_0 0, A[b] < \mathrm{lh}(b)\} & \textit{if such a b exists} \\ \varnothing & \textit{else} \end{cases},$$

$$K_A x := \begin{cases} 0 & \textit{if } A[x] < \mathrm{lh}(x) \\ 1 + \max\{K_A(x*0), K_A(x*1)\} & \textit{else} \end{cases}.$$

In particular, these terms define total functionals in $\mathcal{M}^\omega$ and $\mathcal{C}^\omega$.

Majorants for Howard's Solution

First, note that any term t majorized by a term t^* has at most the complexity of t^* since it can be obtained by bounded search up to t^*. As the complexity of the realizers in theorem 1.78 is not obvious it might be interesting to examine suitable majorants for these realizers:

Theorem 1.79. *The solution of the ND-interpretation of* $\mathsf{WKL}(f)$ *is, provably in* $\mathsf{WE\text{-}HA}^\omega + \mathsf{QF\text{-}AC}^{1,0} + \mathsf{B'}_1$, *majorized as follows:*

$$K^* :=_{(10)2} \lambda A^2, x^0.A^2\mathbf{1} \ \mathsf{maj}_{(10)2} \ K \tag{1}$$

$$B^* :=_{1(2)} \lambda A^2.\mathbf{1}^1 \ \mathsf{maj}_{1(2)} \ B. \tag{2}$$

Again, these terms define total functionals in $\mathcal{M}^\omega$ and $\mathcal{C}^\omega$.

Proof.

(2) Follows directly from the fact:

$$\lambda n^0, x^0.\overline{\mathbf{1}^1}(\mathrm{lh}(x) + n) \ \mathsf{maj}_{1(1)} \ F.$$

(1) According to definition 1.27-(2) suppose A^* majorizes A. We get immediately from the definition of K_A and $A^* \ \mathsf{maj}_2 \ A$

$$\forall s \in \{0,1\}^{A^*\mathbf{1}} \forall x \subseteq s \ K_A x \ \leq_0 \ A^*\mathbf{1} - \mathrm{lh}(x).$$

In other words, for any finite binary sequence x we have:

$$\mathrm{lh}(x) \leq_0 A^*\mathbf{1} \to K_A x \leq_0 A^*\mathbf{1}.$$

Now, suppose $\mathrm{lh}(x) > A^*\mathbf{1}$ for some x. From $A^* \mathrm{maj}_2\, A$ we get $\mathrm{lh}(x) > A[x]$ and we have immediately from the definition of K_A that $K_A x =_0 0 \leq_0 A^*\mathbf{1}$.
So finally, we obtain: $K_A x \leq_0 A^*\mathbf{1}$, what is the same as $A^*\mathbf{1}\, \mathrm{maj}_0\, K_A x$ for all finite binary sequences x.

□

Summary

Above, we discussed at length the WKL principle purposefully following the notation as used by Kohlenbach in [71], which would unfortunately clash with the remainder of this thesis. Therefore, let us here briefly summarize the important results regarding WKL using a more compatible notation. Also, we resort directly to Shoenfield interpretation.

Definition 1.80. $\mathsf{WKL}_\Delta \equiv \forall f \mathsf{WKL}_\Delta(f)$, where

$$\mathsf{WKL}_\Delta(f) \quad :\equiv \quad \exists b^1 \forall k^0\, \widehat{\left(\widehat{f}\right)_g}(\bar{b}k) =_0 0, \text{ with}$$

$$\widehat{f}\, n := \begin{cases} fn & \text{if } fn \neq 0 \lor (\forall k, l(k * l = n \to fk = 0) \land \forall i < \mathrm{lh}(n)\,(n_i \leq 1)) \\ 1^0 & \text{else,} \end{cases}$$

$$f_g n := \begin{cases} fn & \text{if } f(g(\mathrm{lh}(n))) = 0 \land \mathrm{lh}\,(g(\mathrm{lh}(n))) = \mathrm{lh}(n) \\ 0^0 & \text{else,} \end{cases}$$

$$gk := gfk := \begin{cases} \min n \leq \overline{\mathbf{1}^1}k\, (\mathrm{lh}(n) = k \land fn = 0) & \text{if such an } n \text{ exists} \\ 0^0 & \text{else,} \end{cases}$$

where for any given number theoretical function f^1, $\widehat{f}$ assigns a unique characteristic function of a $0/1$-tree[6] and $f_g n$ adds the full subtree if there is no path of length n[7] (this may destroy the tree property of f if present). The function g simply looks for a path of length n and retruns 0 if none exists (otherwise the code of the path itself is returned).

We have: $\widehat{\mathsf{WE\text{-}HA}}^\omega {\restriction}\!\vdash \mathsf{WKL}_\Delta \leftrightarrow \mathsf{WKL}$.
Howard proves in [53] that one can give the realizing functionals for the Sh-interpretation of WKL using only restricted bar recursion and $\mathcal{T}_0$. This proof is discussed in great detail above and we use it to obtain the Sh-interpretation of $\mathsf{WKL}(f)$:

Theorem 1.81 (The Sh-interpretation of WKL_Δ). *The Weak König's lemma for binary trees*

$$\forall f \exists b^1 \forall k^0\, \widehat{\left(\widehat{f}\right)_g}(\bar{b}k) =_0 0$$

is, provably in $\mathsf{WE\text{-}HA}^\omega + \mathsf{B'}_1$, *Sh-interpreted as follows:*

$$\forall f, A\, \exists b^1\, \widehat{\left(\widehat{f}\right)_g}(\bar{b}(Ab)) = 0,$$

---

[6] If f was such a characteristic function already, then it is not modified at all (i.e. we would have $\widehat{f} = f$).
[7] Again, if f defined an infinite tree already, then it is not modified at all (i.e. we would have $f_g = f$).

where b is realized by $b :=_1 B^{\mathsf{WKL}} Af$*:*

$$B^{\mathsf{WKL}}(A,f) := \left[F^{\mathsf{WKL}}\left(A, \widehat{\left(\widehat{f}\right)_g}, K^{\mathsf{WKL}}(A,\varnothing)\right)\right],$$

$$F^{\mathsf{WKL}}(A,f,n) := \begin{cases} \min_0\{b^0 : \mathrm{lh}(b^0) \le n, f(b) =_0 0, A[b] < \mathrm{lh}(b)\} & \text{if such a } b \text{ exists} \\ \varnothing & \text{else} \end{cases}$$

$$K^{\mathsf{WKL}}(A,x) := \begin{cases} 0 & \text{if } A[x] < \mathrm{lh}(x) \\ 1 + \max\{K_A(x*0), K_A(x*1)\} & \text{else} \end{cases}.$$

Note that K^{WKL} *is definable by* B'$_1$.

Proposition 1.82. *The solution of the Sh-interpretation of* WKL(f) *is, provably in* $\widehat{\mathsf{WE\text{-}HA}}^{\omega}\!\upharpoonright$ +B'$_1$*, majorized as follows:*

$$K^* :=_{1(0)(2)} \lambda A^2, x^0.A^2\mathbf{1}\ \mathrm{maj}_{1(0)(2)}\ K^{\mathsf{WKL}}, \quad B^* :=_{1(2)} \lambda A^2.\mathbf{1}^1\ \mathrm{maj}_{1(2)}\ B^{\mathsf{WKL}}.$$

1.17 Complexity classes and bar recursion

The interesting question, apart from the concrete form of the realizers, is how the specific bar recursors and rules for bar recursion affect their complexity. Considerable research was done in the general field of investigating the effect of bar recursion on the complexity of the provably total functions of the underlying system by W. A. Howard, G. Kreisel, H. Luckhardt, H. Schwichtenberg and others. We will mainly use the results of Howard (see [50], [53]), who gave an ordinal analysis of bar recursion of type 0 for the cases in which the bar recursion operator has type level 3 or 4 (here we mean the final type level of the operator, not the type level of its arguments) and studied the effect of recursors of specific types on the complexity in general.

There are several different notions of *primitive recursive* functionals. In connection with the D-interpretation Gödel considered the following class of functionals:

Definition 1.83 (Gödel's $\mathcal{T}$, Gödel 1958)**.** The set theoretic functionals denoted by the closed terms of E-HA$^\omega$/WE-HA$^\omega$ are called *Gödel primitive recursive functionals of finite type.* The quantifier-free term calculus corresponding to WE-HA$^\omega$ is also called Gödel's $\mathcal{T}$.

In addition we define the *α-recursive* functionals:

Definition 1.84 ((unnested) Ordinal Recursion)**.** Let the well-ordering of natural numbers $\prec$ be as in [110]. By $\prec_\alpha$ we mean the restriction of $\prec$ to numbers $n \prec \alpha$, i.e. $a \prec_\alpha b :\equiv a \prec b \prec \alpha$. A function defined by means of a sequence of explicit definitions and the (unnested) *ordinal recursion* on $\prec_\alpha$ (where y' is the successor of y):

$$f^1(\underline{x}, 0) = G^1(\underline{x}) \tag{1}$$

$$f^1(\underline{x}, y') = H^1\left(\underline{x}, y', f(\underline{x}, \theta(\underline{x}, y'))\right), \tag{2}$$

where $\forall \underline{x}, y\ (\theta(\underline{x}, y') \prec_\alpha y' \wedge \theta(\underline{x}, 0) = 0)$, is called an *α-recursive* function. By $<\alpha$-recursive, we mean a β-recursive function with $\beta < \alpha$.

Note that in this section α is used to denote an ordinal as well as to denote a variable, depending on the context.
The class of primitive recursive functionals in the sense of Kleene (see S1-S8 in [60]) is strictly smaller than the class of Gödel primitive recursive functionals. In fact, the primitive recursive functionals in the sense of Kleene are the functionals of pure type defined by the closed terms of $\widehat{\mathsf{WE\text{-}HA}}^{\omega}\restriction / \widehat{\mathsf{E\text{-}HA}}^{\omega}\restriction$ as discussed by Feferman in [25].
For type 1, the Kleene primitive recursive functionals are just the ordinary primitive recursive functions whereas the Gödel primitive recursive functionals of type 1 are the provably total recursive functions of PA, that is $<\epsilon_0$-recursive functions.
To enable a more subtle description for the complexity of terms we introduce the restricted classes of Gödel's $\mathcal{T}$.

Definition 1.85 ($\mathcal{T}_n$)**.** By $\mathcal{T}_n$ we denote the fragment of $\mathcal{T}$ with R_ρ (see definition 1.3) restricted to ρ of type level $\leq n$.

The first interesting question is, how does the Bar Recursion affect the primitive recursive functionals in $\mathcal{T}$. In other words, how does the addition of B (and its defining axioms) to the systems in definition 1.83 above change the class of such functionals.
This was studied in different ways by several researchers. One of the first to publish a partial answer was H. Schwichtenberg:

Proposition 1.86 (Schwichtenberg [111])**.** *For functionals y, z, u in $\mathcal{T}$ of proper types, also the functionals $\mathsf{B}_{0,\rho}yzu$ and $\mathsf{B}_{1,\rho}yzu$ are in $\mathcal{T}$ for arbitrary $\rho \in \mathbf{T}$.*

In other words, the terms of $\mathcal{T}$ are closed under the rule of bar recursion $\mathsf{B}_{1,\rho}$. This result was used by Kohlenbach in [67] to obtain:

Proposition 1.87 (Kohlenbach [67])**.** *Let $t^2[\underline{x}^0, \underline{h}^1]$ be a term of $\mathcal{T}$ containing at most the free variables $\underline{x}$ of type 0 and variables $\underline{h}$ of type level 1. Let z, u, n, and y be the respective arguments of $\mathsf{B}_{0,\tau}$ of appropriate type for arbitrary $\tau \in \mathbf{T}$. Then the functional*

$$\lambda \underline{x}, \underline{h}, z, u, n, y.\mathsf{B}_{0,\tau}(t[\underline{x}, \underline{h}], z, u, n, y)$$

is definable in $\mathcal{T}$ such that $\mathsf{WE\text{-}HA}^{\omega}$ proves its characterizing equations.

Using this result and a normalization argument Kohlenbach proves in [67] also another related result (Proposition 4.1, page 1504) of which we state the following corollary (note that the corollary itself can be concluded also from Howard's results in [53]):

Corollary 1.88. *Up to type level 2, the terms (containing only variables of type level ≤ 1) definable in $\mathcal{T}_1 + \mathsf{B}_{0,1}$ are definable in $\mathcal{T}$ and vice versa.*

We should mention that the proof of 1.86 was based on ordinal analysis using Tait's result given in [120]:

Proposition 1.89 (Tait [120], p. 189–191)**.** *Given an ordinal $\alpha < \epsilon_0$, the α-recursion can be reduced to primitive recursion at higher type.*

For the purpose of this thesis an earlier published result of Parsons is even more interesting:

Definition 1.90. Let $\omega_k(\omega)$ denote $\omega^{\omega^{\cdot^{\cdot^{\omega}}}}\}k$.

Proposition 1.91 (Parsons [103], p. 361). *The same functions are $<\omega_{k+1}(\omega)$-recursive as are functions defined by the closed terms of $\mathcal{T}_k$.*

The search for a finer analysis of bar recursion of type 0 suitable to examine cases where only simple forms of bar recursion are added to $\mathcal{T}_n$ only, rather than to full $\mathcal{T}$, was brought forward by Howard in [53].
For a term t^0, Howard defines its *computation size* as the length of its computation tree allowing nondeterministic contractions as defined in [52] and [53]. The corresponding deterministic contraction for Howard's bar recursor, B_H, is defined as follows (see [53]):

$$\mathsf{B}_H AFGc\mathsf{H} \quad \text{contr} \quad R_0(\lambda a, b.Gc\mathsf{H})(Fc(\lambda u^0.\mathsf{B}_H AFG(c * u))\mathsf{H})(\text{lh}(c)\dot{-}A[c]), \qquad (+)$$

where c is the encoding of a finite sequence of natural numbers $c_0, c_1, \ldots, c_k$ for some k and H stands for $H_1 \cdots H_n$ for $H_i, i \in \{1, \ldots, n\}$, of appropriate type. This corresponds to our definition of the bar recursor $\mathsf{B}_{0,\tau}$, (definition 1.39), where A, F, G, and c correspond to y, u, z, and $\overline{x, n}$ respectively.
For a term t^1 or t^2, there are variables $a_1, a_2, \ldots, a_n$ s.t. $ta_1 \ldots a_n$ has type 0. Howard then defines the computational size of t as the computational size for $ta_1 \ldots a_n$. Only terms with type level at most 2 with free variables of type level at most 1 are considered.
Howard extends $\mathcal{T} + \mathsf{B}_H$ by the terms $\{\alpha, c, t\}$ of the same type as $\mathsf{B}_H A$, where the forming of $\{\alpha, c, t\}$ binds all occurrences of α in t and the proper subterms of $\{\alpha, c, t\}$ consist of all subterms of t. For computation the contraction (+) is replaced by the following four contractions (a subterm αm of $\{\alpha, c, t\}$ can be contracted only when $m < \text{lh}(c)$ and only to c_m):

1. $$\mathsf{B}_H AFGc\mathsf{H} \quad \text{contr} \quad \{\alpha, c, A\alpha\}FGc\mathsf{H},$$
   where α is chosen so as not to be free in A.

2. If t is a numeral $< \text{lh}(c)$ then
   $$\{\alpha, c, t\}FGc\mathsf{H} \quad \text{contr} \quad Gc\mathsf{H}.$$

3. For every t of type 0
   $$\{\alpha, c, t\}FGc\mathsf{H} \quad \text{contr} \quad R_0(\lambda a, b.Gc\mathsf{H})(Fc(\lambda u^0.\{\alpha, c, t\}FG(c * u))\mathsf{H})(\text{lh}(c)\dot{-}t_c),$$
   where t_c denotes the result of substituting $[c]$ for α in t.

4. $$\{\alpha, d, t\}FG(d * n)\mathsf{H} \quad \text{contr} \quad \{\alpha, d * n, t\}FG(d * n)\mathsf{H}.$$

From section 4, "Constructive treatment", of [53] it follows that the function corresponding to a term of computational size α is an α-recursive function and vice-versa.
We will use mainly the following result of Howard:

Proposition 1.92 (Howard [53], p. 23). *Let F, G and t have computation sizes f,g and $size(t)$, respectively, and suppose $\{\alpha, c, t\}$ has type level ≤ 3. Then $\{\alpha, c, t\}FGc$ has computation size ω^{g+f2h}, where $h = \omega size(t) + \omega$.*

In other words proposition 1.92 states that:

> *Any recursor, having (as functional) type level* ≤ 3 *(this includes* $\mathsf{B}_{0,1}$*), applied to ordinal recursive functionals on standard orderings up to* $<\omega_k(\omega)$ *results in an ordinal recursive functional on standard ordering* $<\omega_{k+1}(\omega)$.

The interested reader is encouraged to study the original paper. Here, for brevity, we restrict ourselves just to the simplified reformulation above, using the notation of proposition 1.91, rather than introduce the complete set of definitions and notation used by Howard.

Effective learnability

The Cauchy property of a sequence of real numbers states that in order to know the limit of the sequence up to an approximation 2^{-k} we have to compute the n-th element of the sequence. This n is the most important information we have about the value of the limit of the sequence, unfortunately it is often non-computable in the parameters of the sequence. The paper is about the next best information we may have, an effective bound to the number of mistakes we may do while making assumptions on the value of n. This goal is interesting and the paper contains some relevant results in this direction ...

– anonymous referee about [83], Nov 22, 2012.

Motivation

As mentioned earlier, in this chapter we investigate various levels of (effective) quantitative information on convergence theorems or, more precisely, theorems about Cauchy property of a sequence in a metric space (X,d)

$$\forall k \in \mathbb{N}\, \exists n \in \mathbb{N}\, \forall m, \tilde{m} \geq n \; \left(d(x_m, x_{\tilde{m}}) \leq 2^{-k}\right)$$

and also more general Π_3^0-theorems. The probably weakest information is a rate of metastability, namely a functional Φ (note that $\Phi(k,f)$ is in particular simply a uniform bound for the n.c.i. of the formula above – see Remark 1.21) such that

$$\forall k \in \mathbb{N}\, \forall f : \mathbb{N} \to \mathbb{N}\, \exists n \leq \Phi(k,f)\, \forall i,j \in [n; n+f(n)] \; \left(d(x_i, x_j) \leq 2^{-k}\right).$$

In next chapter we will discuss some general logical metatheorems for strong systems of analysis based on full classical logic, which guarantee the extractability of (sub-)recursive (and highly uniform) rates of metastability and give a case study on such an extraction.

Another such result is the already discussed extraction of a uniform rate of metastability for the strong convergence in the mean ergodic theorem for uniformly convex Banach spaces X carried out in [80]. The also aforementioned recent observation by Avigad and Rute [10] that the analysis in [80] can be used to obtain a simple effective (and also highly uniform) bound on the number of fluctuations, was one of the main inspirations for the article [83]. We base large parts of this chapter on [83]. With our

effective learnability, we will try to give answer to the question whether there are general logical conditions on convergence proofs to guarantee the extractability of effective bounds on fluctuations, which arise naturally.
Of course, the extractability of a computable rate of convergence is a sufficient condition for this, however this topic is well known (see e.g. [71] or [36]). Let us note though, that all such criteria are based on intuitionistic systems, with the use of the law-of-excluded-middle limited to LEM$_\neg$. Moreover, adding an important weak principle of classical logic, the so-called Markov principle (extended to all finite types)

$$\mathrm{M}^\omega \; : \; \neg\neg\exists \underline{x}^{\underline{\rho}} \, \varphi_0(\underline{x}) \to \exists \underline{x}^{\underline{\rho}} \, \varphi_0(\underline{x}),$$

is permissible only if LEM$_\neg$ is weakened to the so-called lesser-limited-omniscience-principle LLPO. This is particularly interesting, since this is one of the standard examples for the fact, that in non-standard systems some amount of the law-of-excluded-middle comes so to say automatically. We discuss such a system in the last chapter.

The smallest step towards classical logic not covered by either of the above (but provable in their union) is Σ^0_1-LEM. Note that already HA+Σ^0_1-LEM allows to prove the Cauchyness of the Specker sequence, primitive recursive monotone decreasing sequences of rational numbers in $[0,1]$ which do not have a computable rate of convergence, while the scenarios above typically include AC (which would make Σ^0_1-LEM significantly stronger). Much related work to this was done in [124, 1, 44, 22, 13, 5, 4, 3]. An important observation is that Σ^0_1-LEM is already the general case. Namely, as far as Π^0_3-theorems go[8], there is no difference in proofs based on full classical logic and proofs using only Σ^0_1-LEM. We discuss this in detail prior to Proposition 2.20, where we formalize and prove this claim.
So in order to have a stronger computational content than simply metastability, but also to cover more than the well studied case when we have a full and computable rate of convergence we have to restrict amount of use of the law-of-excluded-middle somewhere between Σ^0_1-LEM and LEM$_\neg$.
It turns out, that the right amount is to allow Σ^0_1-LEM to be used only fixed many times as an instance. See Theorem 2.11, where we show that from such proofs one can always extract effective (and in fact primitive recursive in the sense of Gödel's T) bounds B, L on the effective learnability of a rate of convergence of the sequence in question. Here B, L are effective functionals (in the parameters of the proved statement). We call L the learner or learning procedure and B the bound on the number of steps along this procedure. We will show that this is a strictly stronger information than a rate of metastability as the latter can be obtained from the lernability by a simple and uniform primitive recursive procedure (in the ordinary sense of Kleene – which we will use in this motivation as default). However, there are primitive recursive Cauchy sequences with a primitive recursive rate of metastability which do not admit any computable bound for the learnability (of a rate of convergence).
We will further show that the B, L learnability of a Cauchy statement demonstrates itself in a particular simple structure of the rate of metastability of the underlying sequence:

$$(L^*(\underline{a}^*) \circ g)^{(B^*(\underline{a}^*))}(0),$$

where $f^{(x)}(0)$ denotes the x-times iteration of the function f starting from 0 (note that the majornts B^*, L^* do not involve the counterfunction). It is precisely this form of a

[8] even with parameters in $\mathbb{N}, \mathbb{N}^{\mathbb{N}}$

rate of metastability that has been observed many times in concrete unwindings in ergodic theory and fixed point theory (see e.g. [8, 80, 81, 73, 77, 78]). For the first time this, until the knowledge of this result rather dazzling, circumstance can be explained in natural terms. This is because we not only connect the underlying logical structure but also give a rather natural condition, which is to be expected to occur.
On the other hand, this makes our results from the next chapter even more interesting. Namely, notable exceptions to this restricted format of metastability are the rates of metastability for the ergodic theorem for odd (and even more general) operators in [107] where Wittmann's proof maps to a nested use of the iteration procedure and for the (weak convergence in the) Baillon nonlinear ergodic theorem in [76] where the proof maps nested use of a bar recursive functional. However, both underlying proofs violate our criterion of a bounded use of Σ_1^0-LEM as defined in 2.11.

Comming back to the original motivation, a bound on the number of fluctuations is a strictly stronger quantitative information still (we give a separating example). Together with the already discussed Specker sequences we get the (w.r.t. effectivity) strict hierarchy of quantitative data for the convergence of Cauchy sequences (with – as quoted at various places above – a plenty of real examples from analysis/ergodic theory as they arise in praxis of proof mining for each level, in particular for the newly defined level 3):

1. A rate of convergence.
2. A bound on the number of approximate fluctuations.
3. The B, L-learnability of a rate of convergence by B-many mind changes by a learning procedure L (see below for a precise definition).
4. A rate of metastability.

2.1 Fluctuations versus effective learnability

To be specific, let us use in the following the language of (intuitionistic) arithmetic in all finite types HA^ω (more precisely the system WE-HA^ω, see [71]) as well as its extension $\mathrm{HA}^\omega[X, \|\cdot\|]$ by an abstract normed space X in the sense of [70, 36, 71] in order to be able to cover also the aforementioned recent applications of proof mining to ergodic and fixed point theory which need this enriched language. Everything we say extends mutatis mutandis also to theories where more conditions on X are prescribed (e.g. X being a uniformly convex or a Hilbert space) and convex subsets C of X being added as well as to metric structures X such as metric, W-hyperbolic and CAT(0) spaces (see [71] for all this). The type of natural numbers $\mathbb{N}$ is usually denoted by 0 while 1 denotes the type of functions $\mathbb{N} \to \mathbb{N}$.
$\mathcal{S}^{\omega,X}$ denotes the full set-theoretic model of these theories over the base types $\mathbb{N}$ and X. Occasionally, we will need the relation 'x^* majorizes x' (short: x^* *maj* x) due to W.A. Howard (for the finite types over $\mathbb{N}$) which is defined in the usual hereditary way by induction on the type of x starting from

$$x^* \; maj_0 \; x \; :\equiv x^* \geq x$$

for x^*, x of type 0 and

$$x^* \; maj_X \; x \; :\equiv x^* \geq \|x\|$$

for x of type X and x^* of type 0.

Remark 2.1. Throughout this chapter, we will use several encodings. We use j for the Cantor pairing function and j_1 and j_2 for the corresponding projections. Moreover, we use $\langle \underline{a} \rangle$ for both

- a surjective primitive recursive encoding of tuples of a given length l, with the corresponding projections $j_1, \ldots, j_l$ and
- a surjective sequence encoding (which then includes the length of the encoded sequence) with primitive recursive functions for length lh, concatenation $*$, and projection $(\cdot)_{(\cdot)}$ (i.e. $(n)_k$ for the $k+1$-th element of a finite sequence encoded by n and 0 for $k \geq lh(n)$,) when there is not danger of confusion we use also the simpler notation n_k). For details see [71].
- for both the tuple and the sequence encoding we assume the coding to be increasing in each component and that $\langle a_0, \ldots, a_{k-1} \rangle \geq a_i$ for $i < k$.

For a specific encoding satisfying these requirements see [71], where the sequence encoding is denoted by $\langle a_0, \ldots, a_{k-1} \rangle$ while the k-tuple encoding is denoted by $\nu^k(a_0, \ldots, a_{k-1})$.
Whether we mean a tuple or a sequence coding should be mostly clear from the context (roughly, we mean tuple encoding whenever the length is fixed and sequence encoding otherwise), but whenever this is relevant we say also explicitly which encoding is meant.

Definition 2.2 (the number of fluctuations)**.** For a sequence $x_{(\cdot)}$ in some metric space (X, d) and an $\epsilon > 0$ let $\mathrm{Fluc}(n, i, j)$ denote that there are n fluctuations whose indexes are encoded into i and j.

$$\begin{aligned}\mathrm{Fluc}(n,i,j) :\equiv \mathrm{Fluc}_{x_{(\cdot)},\epsilon}(n,i,j) :\equiv \quad & \mathrm{lh}(i) = \mathrm{lh}(j) = n \quad \wedge \\ & \forall k < n \ (i_k < j_k) \quad \wedge \\ & \forall k < n-1 \ (j_k \leq i_{k+1}) \quad \wedge \\ & \forall k < n \ (d(x_{i_k}, x_{j_k}) > \epsilon).\end{aligned}$$

We call b a bound on the number of ϵ-fluctuations of $x_{(\cdot)}$, iff

$$\forall n > b \forall i, j \neg \mathrm{Fluc}(n, i, j).$$

We call b effective if it is computable in $\epsilon \in \mathbb{Q}^*_+$ and $x_{(\cdot)}$.

In the Language identification in the limit model for inductive inference, the notion of learnable with an existence of a mind change bound was introduced in the sixties (see e.g. [40]). We define a similar concept in the context of general formal theories like PA^ω. Since we require both the learning procedure and the bound on mind changes to be recursive (effective) we call this property *effective learnability*.

Remark 2.3. The proof-theoretic study of learnability by finitely many (though not necessarily effectively bounded) mind changes in analysis has been initiated by Hayashi (see e.g. [42, 43]) who (with Nakata) established the close relation of this concept to limit computability (see e.g. [44]). The concept of mind change for Cauchy statements is also implicit in section 5.1 of [132] (Proof of Lemma 31.c). Effective learnability concepts for functionals $F : D \to \mathbb{N}^\mathbb{N}$ (with $D \subseteq \mathbb{N}^\mathbb{N}$) have recently been investigated in [46].

On the one hand we would like the learning procedure to be as simple as possible and on the other hand we would like to formalize that it can access as much finite information as is available (at a given learning step). In the case of monotone formulas (which is a rather rich class of statements including, in particular, all Cauchy statements), there is a straightforward answer (see Definition 2.4 and Proposition 2.5) allowing us to simplify the theory of learnability, if we assume monotonicity. We give a more general definition (Definition 2.9), which coincides with Definition 2.4 in the monotone case, a few pages later.

Definition 2.4 ((B, L)-learnable monotone formulas)**.** Consider a Σ_2^0 formula φ with the only parameters $\underline{a}^{\underline{\sigma}}$, i.e.

$$\varphi \equiv \exists n^0 \forall x^0\ \varphi_0(x, n, \underline{a}),$$

which is monotone in n, i.e.

$$\forall n^0\ \forall n' \geq n\ \forall x^0\ \big(\varphi_0\ (x, n, \underline{a}) \to \varphi_0(x, n', \underline{a})\big).$$

We call such a formula φ *(B,L)-learnable*, if there are function(al)s B and L such that the following holds (in the full set-theoretic model $\mathcal{S}^{\omega,X}$):

$$\exists i \leq B(\underline{a})\ \forall x\ \varphi_0(x, c_i, \underline{a}),$$

where

$$c_0 := 0,$$
$$c_{i+1} := \begin{cases} L(x, \underline{a}), & \text{for the } x \text{ with } \neg\varphi_0(x, c_i, \underline{a}) \wedge \forall y < x\ \varphi_0(y, c_i, \underline{a}) \text{ if it exists} \\ c_i, & \text{else.} \end{cases}$$

We call such a φ *effectively learnable (with effectively bounded many mind changes)* if it is (B, L)-learnable with computable functionals B and L.[9]

In Proposition 2.33 we will construct a φ which is true for all parameters $\underline{a} \in \mathbb{N}^{\mathbb{N}}$ but which is not (B, L)-learnable with computable B, L.

This definition is very intuitive in the sense that it formalizes the concept of an (effective) learning process L which learns the witness in an effectively bounded number of attempts in a very straightforward way.

Moreover, this definition allows the learner, i.e. the function L to use the least amount of non-computable information possible, namely only the smallest counterexample to the learners last candidate for the witness. Nevertheless, we will show that this amount of information is, in a sense, already exhaustive. More precisely, we have the following (we use in the rest of this section a surjective sequence coding denoted by $\langle \cdots \rangle$ with primitive recursive functions $lh, *, (n)_k$ as discussed in Remark 2.1):

Proposition 2.5. *Consider a monotone formula φ as above. Suppose there are B and L' s.t.*

$$\exists i \leq B(\underline{a})\ \forall x\ \varphi_0(x, c'_i, \underline{a}),$$

[9]Note that c_i is the i-th attempt to produce a candidate for a valid n, while B is a Bound on the number of such attempts produced before a valid candidate is Learned by the procedure L.

where this time L' can access all reasonable information, i.e.

$$c'_0 := 0,$$

$$c'_{i+1} := \begin{cases} L'(\langle x_0,\dots,x_i\rangle, \langle c'_0,\dots,c'_i\rangle, \underline{a}), & \text{for those } x_j, c'_j, j \le i \text{ with} \\ & \neg\varphi_0(x_j, c'_j, \underline{a}) \wedge \forall y < x_j\ \varphi_0(y, c'_j, \underline{a}) \\ & \text{if each exists,} \\ c'_i, & \text{else.} \end{cases}$$

Then φ is (B, L)-learnable in the original sense (as defined in Definition 2.4), where L is primitive recursively definable in B, L' and the characteristic function of φ_0 (and so, in particular as $\xi(B, L')$ for a closed term ξ of the system at hand).

Proof. W.l.o.g we can assume that there is a $j \le B(\underline{a})$ s.t. $\forall i < j\ (c'_{i+1} > c'_i)$ and $\forall i \ge j\ (c'_{i+1} = c'_i)$ (we can actually primitive recursively define a learner which satisfies this property whenever we have B and L' as above). We set

$$L(x,\underline{a}) := L(\langle\rangle,\langle\rangle,x,\underline{a}) := \begin{cases} 0, & \text{if} \varphi_0(x,0,\underline{a}), \\ L(\langle X(x,0,\underline{a})\rangle, \langle 0\rangle, x, \underline{a}), & \text{else,} \end{cases}$$

$$L(\underbrace{\langle x_0,\dots,x_i\rangle}_{\underline{x}}, \underbrace{\langle c_0,\dots,c_i\rangle}_{\underline{c}}, x, \underline{a}) := \begin{cases} c_i, & \text{if} \bigvee_{j\le i} \varphi_0(x_j,c_j,\underline{a}) \vee i \ge B(\underline{a}), \\ l' := L'(\langle\underline{x}\rangle, \langle\underline{c}\rangle, \underline{a}), & \text{if} x = x_i \vee \varphi_0(x,l',\underline{a}), \\ & \bigwedge_{j\le i} \neg\varphi_0(x_j,c_j,\underline{a}) \wedge i < B(\underline{a}), \\ L(\langle \underline{x}, X(x,l',\underline{a})\rangle, \langle \underline{c}, l'\rangle, x, \underline{a}) & \text{else,} \end{cases}$$

where $X(x,c,\underline{a}) := \min\{x' \le x : \neg\varphi_0(x',c,\underline{a})\}$ (or x if there is no such x').
We show by induction on i that $\forall i\ (c'_i = c_i)$. This is obvious for $i = 0$, moreover if $\forall x\ \varphi_0(x, c'_i, \underline{a})$ then also $\forall x\ \varphi_0(x, c_i, \underline{a})$ and so $c'_{(\cdot)} = c_{(\cdot)}$ (both) by the induction hypothesis $\forall j \le i (c'_j = c_j)$.
Otherwise, we have the smallest counterexample x'_i to c'_i, and since by our hypothesis $c'_i = c_i$ we have also $x_i = x'_i$ for the smallest counterexample to c_i (note that, by the (B, L')-learnability of φ, we have $i < B(\underline{a})$). So, by the monotonicity of φ, we obtain for all $j < i$ that $x'_j \le x_i$, so $X(x_i, c'_j, \underline{a}) = x'_j$ and by definition of L we get in total that

$$\begin{aligned} c_{i+1} = L(x_i,\underline{a}) &= L(\langle X(x_i,0,\underline{a})\rangle, \langle 0\rangle, x_i, \underline{a}) \\ &= L(\langle x'_0\rangle, \langle 0\rangle, x_i, \underline{a}) \\ &= L(\langle x'_0, X(x_i, L'(\langle x'_0\rangle, \langle 0\rangle, \underline{a}), \underline{a})\rangle, \langle 0, L'(\langle x'_0\rangle, \langle 0\rangle, \underline{a})\rangle, x_i, \underline{a}) \\ &= L(\langle x'_0, X(x_i, c'_1, \underline{a})\rangle, \langle 0, c'_1\rangle, x_i, \underline{a}) \\ &= L(\langle x'_0, x'_1\rangle, \langle 0, c'_1\rangle, x_i, \underline{a}) \\ &= \dots \\ &= L'(\langle\underline{x}'\rangle, \langle\underline{c}'\rangle, \underline{a}) = c'_{i+1}. \end{aligned}$$

□

So from now on we will simply use L in the form which suits us best.

Speaking of a Cauchy sequence $a_{(\cdot)}$, we would say that it has an effectively learnable rate of convergence, if there is a recursive computation for a bound b from the sequence $a_{(\cdot)}$ (resp. the parameters used in defining $a_{(\cdot)}$) and an $\epsilon > 0$, such that there is a procedure to learn an ϵ-Cauchy point with at most b computable corrections (computable in a counterexample x, which in turn may not be computable itself!).

Remark 2.6. In Definition 2.4, even the condition that x is the smallest counterexample, i.e. $\forall y < x\, \varphi_0(y, c_i, \underline{a})$, is not really necessary. Of course, in such case, for a given learner, the sequence $c_{(\cdot)}$ is not unique and we need to specify what actually is bounded by B. Fortunately, there are only two natural options. Either we say that B is a bound on any sequence $c_{(\cdot)}$ (i.e. B is independent on the choice of the counterexamples) or we say that B is a bound for some sequence $c_{(\cdot)}$ (i.e. for at least one suitable choice of counterexamples). It seems rather obvious that the second option makes little sense, since if there was any bound and learner at all, then $B(\underline{a}) = 1$ would be a correct bound for the same learner as well (simply by choosing the right counterexample as the first input). Moreover, a definition in the new sense that B is a bound on any sequence $c_{(\cdot)}$ (any choice of $x_{(\cdot)}$) would be equivalent to Definition 2.4.

- Any given bound B for all such sequences is obviously, in particular, a bound for the one we used in the original definition. Therefore any formula (B, L)-learnable in the new sense is, in particular, (B, L)-learnable in the old sense.
- On the other hand, given B and L satisfying our original definition, we could modify L to L' in such a way, that it actually looks for the smallest counterexample and uses that for its computation, assuring that we in fact generate the same sequence $c_{(\cdot)}$ after all (e.g. set $L'(x, c, \underline{a}) := L(\min\{x' \leq x : \neg\varphi_0(x', c, \underline{a}) \wedge \forall y < x'\, \varphi_0(y, c, \underline{a})\}, c, \underline{a})$ if such an x' exists and $L'(x, c, \underline{a}) := L(x, c, \underline{a})$ otherwise). In other words, any formula (B, L)-learnable in the old sense, is (B, L')-learnable in the new sense.

As far as monotone formulas are concerned, we have yet another nice property.

Proposition 2.7. *A monotone Σ^0_2-formula φ (see also Definition 2.4) that is (B, L)-learnable (uniformly in the parameters $\underline{a}$) is also (B^*, L^*)-learnable (uniformly in majorants $\underline{a}^*$ of $\underline{a}$) for any majorants B^*,L^* of B, L, i.e. (in $\mathcal{S}^{\omega,X}$)*

$$\forall \underline{a}^*, \underline{a}(\underline{a}^* \text{ maj } \underline{a} \to \exists i \leq B^*(\underline{a}^*)\, \forall x^0\, \varphi_0(x, c^*_i, \underline{a})),$$

where

$$c^*_0 := 0,$$

$$c^*_{i+1} := \begin{cases} L^*(x, \underline{a}^*), & \text{for the } x \text{ with } \neg\varphi_0(x, c^*_i, \underline{a}) \wedge \forall y < x\, \varphi_0(y, c^*_i, \underline{a}) \text{ if it exists} \\ c^*_i, & \text{else.} \end{cases}$$

Proof. Note that $B^*, L^*, \underline{a}^*$ *maj* $B, L, \underline{a}$ implies that

$$B^*(\underline{a}^*) \geq B(\underline{a}) \wedge \forall x^0, y^0(x \geq y \to L^*(x, \underline{a}^*) \geq L(y, \underline{a}))).$$

Now assume that $c^*_i \geq c_i$. Then by the monotonicity of φ we have for all x

$$\neg\varphi_0(x, c^*_i, \underline{a}) \to \neg\varphi_0(x, c_i, \underline{a})$$

and so the smallest counterexample x_i to c_i is smaller (or equal) than the smallest counterexample x^*_i to c^*_i and so $c^*_{i+1} = L^*(x^*_i, \underline{a}^*) \geq L(x_i, \underline{a}) = c_{i+1}$. Inductively, we get $c^*_i \geq c_i$ for all $i \leq B(\underline{a})$. □

Remark 2.8. If the parameters $\underline{a}$ have all types of degree ≤ 1, then φ in the above proposition is learnable in B^*, L^* uniformly in $\underline{a}$ since a^M *maj* a^1, where $a^M(n) := \max\{a(i) : i \leq n.\}$.
In this sense, we can extend the term effectively learnable as follows. A monotone Σ^0_2-formula $\varphi(\underline{a})$ is *effectively learnable with finitely many mind changes uniformly in majorants $\underline{a}^*$ of the parameters $\underline{a}$* if it is (B^*, L^*)-learnable (uniformly in majorants $\underline{a}^*$ of the parameters $\underline{a}$ by computable functionals B^*, L^* and all elements of $\underline{a}^*$ are of type level at most one). Note that this means that in the System $\mathsf{HA}^\omega[X, \|\cdot\|]$, $\underline{a}$ could include parameters of types like $X, \mathbb{N} \to X, X \to X$.

There are several ways to generalize our learnability definition. Of course one can drop the monotonicity condition, but we can also allow higher or abstract types for n and x (for x we would need to consider all sequences $c_{(\cdot)}$ since we can provide only a counterexample x, not the smallest counterexample x – see also Remark 2.6). The question here is, what kind of information we do allow the learning function(al) L to access. At the moment, it seems that there is not such a nice and definitive answer as in the monotone case. However, we will stick with a (not necessarily unique) definition (see Definition 2.9), which

1. generalizes Definition 2.4, i.e. if a monotone formula (assuming the bound variables to be of type 0) is learnable according to our new definition, it is also learnable in the sense of Definition 2.4 and vice-versa,

2. while keeping the arguments of the learner L simple, still is equivalent to the case where the learner has access to the full finitary information in the sense of Proposition 2.5 (see Remark 2.10)

3. fits very nicely into the hierarchy of different concepts for computational information (see Proposition 2.16),

4. makes effective learnability guaranteed by very clear logical conditions on the provability of the learned formula (see Theorem 2.11).

The second point seems a very natural requirement and is the cause for the main difference to the monotone case, which is that in Definition 2.9 the learning process may depend on a whole tuple of all counterexamples used so far, rather than only on the last one (last in the sense of number of guesses/candidates, not the index of the counterexample).
Although we do not treat the case of learnability for higher type objects in this thesis, the following definition easily applies to this case as well and, therefore, is written in this generality:

Definition 2.9 ((B,L)-learnability for general (not necessarily monotone) formulas). Consider an $\exists\forall$ formula φ with the only parameters $\underline{a}^{\underline{\sigma}}$, i.e.

$$\varphi \equiv \exists n^\rho \forall x^0 \; \varphi_0(x, n, \underline{a}).$$

We call such a formula φ *(B, L)-learnable*, if there are function(al)s B and L such that the following holds:

$$\exists i \leq B(\underline{a}) \; \forall x \; \varphi_0(x, c_i, \underline{a}),$$

where

$$c_0 := 0^\rho,$$

$$c_{i+1} := \begin{cases} L(\langle x_0,\ldots,x_i\rangle,\underline{a}), & \text{for those } x_j, j \le i \text{ with} \\ & \neg\varphi_0(x_j,c_j,\underline{a}) \wedge \forall y < x_j\, \varphi_0(y,c_j,\underline{a}) \\ & \text{if each exists,} \\ c_i, & \text{else.} \end{cases}$$

We call such a φ *effectively learnable*, if it is (B,L)-learnable, σ_i and ρ have type level at most one, and B and L are computable.

Remark 2.10. Again, this definition already captures (so to say in a primitive recursive way) the case where the learner could access the previous values of $c_{(\cdot)}$ as well. Simply consider

$$L(\overbrace{\langle x_0,\ldots,x_i\rangle}^{\underline{x}:=},\underline{a}) :=$$
$$L'(\langle\underline{x}\rangle,\langle \underbrace{L'(\langle x_0\rangle,\langle 0\rangle,\underline{a})}_{c_1':=},\underbrace{L'(\langle x_0,x_1\rangle,\langle 0,c_1'\rangle,\underline{a})}_{c_2':=},\ldots,L'(\langle\underline{x}\rangle,\langle 0,c_1',\ldots,c_i'\rangle,\underline{a})\rangle,\underline{a}).$$

Of course, one could consider weaker concepts, like a learner which can access only the last counterexample (as in the monotone case). We considered also a learner of the kind $L'(x,c,\underline{a})$ (i.e. a learner who is allowed to use in addition only the lastly learned solution candidate), which doesn't seem to be equivalent to any of the other two concepts.

Let us make the properties of our learnability definition discussed above more transparent by proving the following results which we first briefly motivate: as mentioned already in the introduction (and proved further below in section 3), any classical proof (in a suitable formal system) of a Cauchy statement $\varphi(k) := \exists n \forall i,j \ge n\, (d(x_i,x_j) < 2^{-k})$ can be reformulated to use classical logic only up to $\forall l^0(\Sigma^0_1\text{-LEM}(s(l,k)))$ (for some closed term s), i.e.

$$(a)\ \forall k^0\ (\forall l^0\,(\Sigma^0_1\text{-LEM}(s(l,k))) \to \varphi(k))$$

follows intuitionistically. This, in particular, applies to the example from section 2.3 of a computable Cauchy sequence in $\mathbb{R}$ which does not possess any computable learnability bounds B, L. So in order to be able to extract such computable data B, L, the Cauchy proof has to be further restricted, namely, to the situation where the proof implicitly contains a bound $t(k)$ on $\forall l^0$, i.e. on the instances of $\Sigma^0_1\text{-LEM}(s(l,k))$ used. This is guaranteed (as we will see) when (a) is strengthened to the (only noneffectively equivalent) form

$$(b)\ \forall k^0\, \exists l^0\ (\forall m \le_0 l\,(\Sigma^0_1\text{-LEM}(s(m,k))) \to \varphi(k)).$$

Then from a (semi-intuitionistic) proof of (b) one can extract a term $B(k)$ computing l and two further terms which allow one to build a learning procedure L so that φ is (B,L)-learnable. In the following $\mathrm{IP}^\omega_\forall$ denotes the independence-of-premise principle for universal premises in all finite types:

$$\mathrm{IP}^\omega_\forall :\ \ (\forall \underline{x}\, A_0(\underline{x}) \to \exists y\, B(y)) \to \exists y\, (\forall \underline{x}\, A_0(\underline{x}) \to B(y)),$$

where A_0 is a quantifier-free formula and $\underline{x}, y$ are variables of arbitrary types.

Theorem 2.11 (Bounded Σ^0_1-LEM guarantees learnability). *Given that*

$$\mathsf{HA}^\omega[X,\|\cdot\|] + \mathrm{AC}{+}\mathrm{M}^\omega{+}\mathrm{IP}^\omega_\forall \vdash \forall \underline{a}\, \exists l^0 \left(\forall m \leq_0 l \exists u^0 \forall v^0 \left(\psi_0(u,m,\underline{a}) \vee \neg\psi_0(v,m,\underline{a})\right) \to \exists n^0 \forall x^0 \varphi_0(x,n,\underline{a})\right),$$

where φ_0, ψ_0 are quantifier-free formulas (containing at most the parameters $\underline{a}$ free), then

$$\exists n \forall x \varphi_0(x,n,\underline{a})$$

is (valid in $\mathcal{S}^{\omega,X}$) (B,L)-learnable (in the sense of Definition 2.9 and, for monotone formulas, in the sense of Definition 2.4) by functionals given by closed terms of $\mathsf{HA}^\omega[X,\|\cdot\|]$.
To B, L one can construct majorants B^*, L^* given by closed terms of HA^ω such that if

$$\exists n \forall x \varphi_0(x,n,\underline{a})$$

is monotone (as in Definition 2.4), then it is even learnable in B^*, L^* uniformly in majorants $\underline{a}^*$ of the parameters $\underline{a}$.

Proof. Suppose that

$$\mathsf{HA}^\omega[X,\|\cdot\|] + \mathrm{AC}{+}\mathrm{M}^\omega{+}\mathrm{IP}^\omega_\forall \vdash \forall \underline{a} \exists l^0 \left(\forall m \leq l \exists u \forall v \left(\psi_0(u,m,\underline{a}) \vee \neg\psi_0(v,m,\underline{a})\right) \to \exists n \forall x \varphi_0(x,n,\underline{a})\right).$$

Then by the soundness of the Gödel functional ('Dialectica') interpretation for $\mathsf{HA}^\omega[X,\|\cdot\|] + \mathrm{AC}{+}\mathrm{M}^\omega{+}\mathrm{IP}^\omega_\forall$ (see [71]) we obtain that (note that since we do not need bar recursion to interpret $\mathsf{HA}^\omega[X,\|\cdot\|]$ we do not have to go through the model of strongly majorizable functionals and so do not need to assume any smallness condition on the types of $\underline{a}$ to pass to $\mathcal{S}^{\omega,X}$)

$$\mathcal{S}^{\omega,X} \models \exists l, V, N \forall U, x \left(\forall m \leq l (\psi_0(Um,m,\underline{a}) \vee \neg\psi_0(VxU,m,\underline{a})) \to \varphi_0(x,N(U),\underline{a})\right).$$

where '$\exists l, V, N$' is witnessed (uniformly in $\underline{a}$) by closed terms t, s_V, s_N of $\mathsf{HA}^\omega[X,\|\cdot\|]$. The result when the terms s_V, s_N are applied to $\underline{a}$, we conveniently name V and N.
To show the learnability, let $U_{\underline{v}}$ (where $\underline{v}$ is a $t(\underline{a})$-tuple) denote the function

$$U_{\underline{v}}(i) := \begin{cases} v_i & \text{if } i < t(\underline{a}), \\ 0 & \text{else,} \end{cases}$$

set $B(\underline{a}) := t(\underline{a})$ and define L in N and V via a sequence of $t(\underline{a})$-tuples $\underline{v}^{(\cdot)}$. More precisely to compute $L(\underbrace{\langle x_0, x_1, \ldots, x_i\rangle}_{\underline{x}:=}, \underline{a})$ for some i we need to define the tuples $\underline{v}^{(0)}, \ldots, \underline{v}^{(i)}$ as follows.

$\underline{v}^0$ Set $\underline{v}^0 := 0,\ldots,0$ and $c_1 := L(\langle x_0\rangle, \underline{a}) := N(U_{\underline{v}^0})$.

$\underline{v}^1$ If $\forall x \varphi_0(x, c_1, \underline{a})$ holds, then there is nothing to be done[10]. Otherwise, we have in particular (provided that x_1 is the minimal counterexample)

$$\exists m \leq t(\underline{a}) (\neg\psi_0(U_{\underline{v}^0} m, m, \underline{a}) \wedge \psi_0(V x_1 (U_{\underline{v}^0}), m, \underline{a}))$$

[10]Of course this is undecidable, however the conclusion discussed next is. In this sense if the conclusion is wrong for the x_1 given as input to L, we can simply set $\underline{v}^1 = \underline{v}^0$ and $L(\langle x_0, x_1\rangle, \underline{a}) := c_1$ (or even 0 for all that it matters).

and so we can denote the least such an m by m_0 (put $m_0 := 0$ in case such an m does not exist) and define $\underline{v}^1$ as $\underline{v}^0$ except that we set $v^1_{m_0} := Vx_1(U_{\underline{v}^0})$. Furthermore, we set $c_2 = L(\langle x_0, x_1\rangle, \underline{a}) := N(U_{\underline{v}^1})$. Note that we have

$$\neg\psi_0(U_{\underline{v}^0} m_0, m_0, \underline{a}) \wedge \psi_0(v^1_{m_0}, m_0, \underline{a}). \tag{v0}$$

$\underline{v}^2$ Now, if $\forall x \varphi_0(x, c_2, \underline{a})$ then we are finished. Otherwise, similarly as before we have

$$\exists m \le t(\underline{a})(\neg\psi_0(U_{\underline{v}^1} m, m, \underline{a}) \wedge \psi_0(Vx_2(U_{\underline{v}^1}), m, \underline{a}))$$

and we can denote the least such an m by m_1 and define $\underline{v}^2$ as $\underline{v}^1$ except that we set $v^2_{m_1} := Vx_2(U_{\underline{v}^1})$. As before this means that

$$\neg\psi_0(U_{\underline{v}^1} m_1, m_1, \underline{a}) \wedge \psi_0(v^2_{m_1}, m_1, \underline{a}), \tag{v1}$$

so in particular we obtain that $m_1 \neq m_0$ by (v0) as $U_{\underline{v}^1} m_1 = v^1_{m_1}$. We set $c_3 = L(\langle x_0, x_1, x_2\rangle, \underline{a}) := N(U_{\underline{v}^2})$ and continue.

$\underline{v}^3$ Again, if $\forall x \varphi_0(x, c_3, \underline{a})$ then we are finished. Otherwise, as before, we have that

$$\exists m \le t(\underline{a})(\neg\psi_0(U_{\underline{v}^2} m, m, \underline{a}) \wedge \psi_0(Vx_3(U_{\underline{v}^2}), m, \underline{a}))$$

and we can denote the least such an m by m_2 and define $\underline{v}^3$ as $\underline{v}^2$ except that we set $v^3_{m_2} := Vx_3(U_{\underline{v}^2})$. As before this means that

$$\neg\psi_0(U_{\underline{v}^2} m_2, m_2, \underline{a}) \wedge \psi_0(v^3_{m_2}, m_2, \underline{a}), \tag{v2}$$

so in particular we obtain that $m_2 \neq m_1$ by (v1). Moreover, by (v0) and (v2) we have $m_2 \neq m_0$, since from $m_1 \neq m_0$ follows that $v^2_{m_0} = v^1_{m_0}$. We set $c_4 = L(\langle x_0, x_1, x_2, x_3\rangle, \underline{a}) := N(U_{\underline{v}^3})$ and continue.

$\underline{v}^{n+1}$ Finally, in general assume that for some n we have that $\forall i < n \forall j < i\, m_i \neq m_j$ and $\forall i \le n+1 \neg\forall x \varphi_0(x, c_i, \underline{a})$. Then we have also that

$$\forall i < n(\neg\psi_0(U_{\underline{v}^i} m_i, m_i, \underline{a}) \wedge \psi_0(Vx_{i+2}(U_{\underline{v}^{i+1}}), m_i, \underline{a})). \tag{vi}$$

As usual we have in particular that

$$\exists m \le t(\underline{a})(\neg\psi_0(U_{\underline{v}^n} m, m, \underline{a}) \wedge \psi_0(Vx_{n+1}(U_{\underline{v}^n}), m, \underline{a}))$$

and we can denote the least such an m by m_n and define $\underline{v}^{n+1}$ as $\underline{v}^n$ except that we set $v^{n+1}_{m_n} := Vx_{n+1}(U_{\underline{v}^n})$. As before this means that

$$\neg\psi_0(U_{\underline{v}^n} m_n, m_n, \underline{a}) \wedge \psi_0(v^{n+1}_{m_n}, m_n, \underline{a}). \tag{vn}$$

From $\forall 0 < i < n\ (m_0 \neq m_i)$ it follows that $\forall 0 < i < n\ (v^n_{m_0} = v^i_{m_0})$. Assume that $m_n = m_0$, then

$$U_{\underline{v}^n} m_n = v^n_{m_n} = v^n_{m_0} = v^1_{m_0}$$

and we obtain a contradiction as $\neg\psi_0(v^1_{m_0}, m_0, \underline{a})$ follows from (vi) and $\psi_0(v^1_{m_0}, m_0, \underline{a})$ follows from (vn). This shows that $m_n \neq m_0$, similarly one shows that

$$\forall i < n\ (m_n \neq m_i).$$

As usual, we set $c_{n+2} = L(\langle x_0, \ldots, x_{n+1}\rangle, \underline{a}) := N(U_{\underline{v}^{n+1}})$.

This leads to the following definition of L:

$$L(x,\underline{a}) := L(\underbrace{\langle x_0, x_1, \ldots, x_i \rangle}_{\underline{x}:=}, \underline{a}) := N(U_{\underline{v}^i}).$$

Note that since N and U are total, so is L. Moreover, if the values of i, $\underline{x}$, and c_i satisfy the conditions from Proposition 2.5, then L behaves as described above.
Finally, since there can be only $t(\underline{a})$ many different m_i's, it can happen at most $t(\underline{a})$ many times that $\forall x \varphi_0(x, c_i, \underline{a})$ does not hold, where c_i is defined as in Definition 2.9 with L as above. Hence φ is (B,L)-learnable in the sense of Definition 2.9 and hence - for monotone formulas - also $(B,\tilde{L})$-learnable for some $\tilde{L}$ primitive recursive in B, L (and φ_0) by Proposition 2.5.
The second claim follows from the fact that t, s_V, s_v have majorants t^*, s_V^*, s_N^* given by closed terms of HA^ω (see [71]) which then yield majorants B^*, L^* of B, L. Now apply Proposition 2.7. □

Remark 2.12. Assume that φ_0 in the theorem comes from a Cauchy statement

$$j_1(x), j_2(x) \geq n \to \|\widehat{a_{j_1(x)} - a_{j_2(x)}}\|(k+1) \leq_{\mathbb{Q}} 2^{-k},$$

where – referring to the representation of real numbers by number theoretic functions f representing fast Cauchy sequences of rationals – $\widehat{f}(k+1)$ is a 2^{-k-1}-rational approximation to f (see [71] for details). Then φ is monotone and a counterexample x to n satisfies $x \geq n$ (using that for the Cantor pairing function $x \geq j_i(x)$). Assume also that ψ_0 is monotone in u (which always can be arranged by taking $\psi_0'(u, m, \underline{a}) :\equiv \exists \tilde{u} \leq u\, \psi_0(u, m, \underline{a})$).
Then the complicated iteration used in defining L can be avoided by taking simply

$$L^*(\langle x_0, \ldots, x_i \rangle, \underline{a}^*) := N^*_{\underline{a}^*}(\lambda k. V^*_{\underline{a}^*}(x_i, \lambda n. x_i)),$$

where $\lambda \underline{a}^*.N^*_{\underline{a}^*}, \lambda \underline{a}^*.V^*_{\underline{a}^*}$ are majorants of

$$\tilde{N}_{\underline{a}}(f) := \max\left\{\max\left\{N(v^0 * 0) : lh(v) = t\underline{a} \wedge \forall l \leq t\underline{a}(v_l \leq f(l))\right\}, f(l) : l \leq t\underline{a}\right\}$$

and

$$\tilde{V}_{\underline{a}}(x,f) := \max\left\{\max\left\{V(x, v^0 * 0) : lh(v) = t\underline{a} \wedge \forall l \leq t\underline{a}(v_l \leq f(l))\right\}, f(l) : l \leq t\underline{a}\right\}$$

with N, V, t as in Theorem 2.11. Note that with N, V also $\tilde{N}, \tilde{V}$ satisfy the claim in the proof and that L^* (for counterexamples $x_0, \ldots, x_i$) is an upper bound for the L defined in terms of $\tilde{N}, \tilde{V}$ as an elementary calculation shows (using that – by the form of φ_0 – a counterexample x to n has to satisfy $x \geq n$). By monotonicity, φ is then also (L^*, B^*)-learnable (uniformly in majorants $\underline{a}^*$ of $\underline{a}$) where B^* is some majorant of B.

Remark 2.13. The theorem remains valid if arbitrary $\mathcal{S}^{\omega,X}$-true purely universal sentences are added as axioms to $\mathrm{HA}^\omega[X, \|\cdot\|]$. The part about the majorizing terms B^*, L^* even remains valid – using monotone functional interpretation – if one adds sentences of the form $\Delta :\equiv \forall a^\delta \exists b \leq_\rho sa \forall c^\tau F_0(a,b,c)$ with quantifier-free F_0 and closed s as axioms which covers the case of the binary ('weak') König's lemma WKL (which together with AC even implies König's lemma KL); see [71] for extensive details on all this.

Using the representation of real numbers from [71], each sequence of type $0 \to 1$ can be viewed as a name of a sequence (a_n) of reals. Now define $\tilde{a}_n := \max_{\mathbb{R}} \left(0, \min_{i \leq n} a_i\right)$. Let $\mathrm{PCM}_{ar}(a_n)$ denote the statement that the monotone decreasing sequence $(\tilde{a}_n)$ in $[0,1]$ is Cauchy (see [71] for details)

$$\mathrm{PCM}_{ar}(a_n) : \forall k \in \mathbb{N} \underbrace{\exists n \in \mathbb{N}\, \forall m \geq n \left(|\tilde{a}_m - \tilde{a}_n| \leq 2^{-k}\right)}_{\mathrm{PCM}_{ar}((a_n),k):\equiv}$$

(if (a_n) is already a decreasing sequence in $[0,1]$, then $(\tilde{a}_n) = (a_n)$). The usual classical proof of $\mathrm{PCM}(a_n)$ uses Σ^0_2-DNE, but it can be converted into a proof that only needs the weaker Σ^0_1-LEM (see Proposition 2.20 below). In [124], an explicit such proof is constructed exhibiting a concrete sequence of instances of Σ^0_1-LEM sufficient for this. From this proof one can read off the following even more detailed fact:

Proposition 2.14. *There is a primitive recursive functional (in the ordinary sense) Φ such that (using the Cantor pairing function)*

$$\mathrm{HA}^\omega \vdash \forall a^{0\to 1}_{(\cdot)}, k^0 \left(\forall m \leq j(2^k - 1, 2^k)\, \Sigma^0_1\text{-LEM}(\Phi(a_{(\cdot)}, m) \to \mathrm{PCM}_{ar}(a_{(\cdot)}, k)\right).$$

Proof. The crucial step in Toftdal's proof in [124] is to show by induction on k that

$$\forall k \exists i \in \{1, \ldots, 2^k\} \exists n \forall m \left(\frac{i-1}{2^k} \leq \tilde{a}_{n+m} \leq \frac{i}{2^k}\right),$$

where in the induction step Σ^0_1-LEM is used in the form (note that, based on our representation of real numbers, $<_{\mathbb{R}} \in \Sigma^0_1$)

$$\exists n \left(\tilde{a}_n < \frac{2i-1}{2^{k+1}}\right) \vee \neg \exists n \left(\tilde{a}_n < \frac{2i-1}{2^{k+1}}\right).$$

So to establish $\mathrm{PCM}_{ar}(a_{(\cdot)}, k)$ one needs only the instances

$$\exists n \left(\tilde{a}_n < \frac{i}{2^l}\right) \vee \neg \exists n \left(\tilde{a}_n < \frac{i}{2^l}\right)$$

for $i \leq l - 1$ and $l \leq 2^k$, i.e. the codes $j(i,l)$ of the instances used can be bounded by $t(k) := j(2^k - 1, 2^k)$. The construction of Φ is clear. □

While the usual classical proof of PCM_{ar} only needs Σ^0_1-induction (but Σ^0_2-DNE), the above proof due to Toftdal needs an instance of the Σ^0_2-induction rule (Σ^0_2-IR) which, apparently, is the price to be paid for using only Σ^0_1-LEM (instead of Σ^0_2-DNE). Classically, Σ^0_2-IR is quite strong and proves (relative to PRA) the same Π^0_3-sentences as Π^0_2-IA (see e.g. [113][Theorem 3.11]) and so, in particular, the totality of the Ackermann function. In our intuitionistic context, however, it is weak and the functional interpretation used (without negative translation!) in the proof of Theorem 2.11 (and the corollary below) to extract B, L solves Σ^0_2-IR using only ordinary primitive recursion in the form of R_0. In fact, one can also show that Π^0_1-CP is sufficient (see [83]).

As a corollary we obtain that Theorem 2.11 also holds with the original assumption being replaced by $\mathrm{PCM}_{ar}(s(\underline{a}, l), t(\underline{a}))$, where $s(\underline{a}, l)^{0\to 1}$ represents for each $l \in \mathbb{N}$ some sequence of reals defined by a closed term s in $\underline{a}$.

Corollary 2.15. *Given that*

$$\begin{aligned}&\mathrm{HA}^\omega[X,\|\cdot\|]+\mathrm{AC}+\mathrm{M}^\omega+\mathrm{IP}^\omega_\forall \vdash\\ &\qquad \forall \underline{a}\,\exists k^0, l^0\,\big(\forall m\le l\,\mathrm{PCM}_{ar}(s(\underline{a},m),k)\to \exists n^0\forall x^0\,\varphi_0(x,n,\underline{a})\big),\end{aligned}$$

where s is a closed term and φ_0 as in Theorem 2.11, then

$$\exists n^0\forall x\,\varphi_0(x,n,\underline{a})$$

is (valid in $\mathcal{S}^{\omega,X}$) (B,L)-learnable (uniformly in $\underline{a}$) by functionals given by closed terms of the system $\mathsf{HA}^\omega[X,\|\cdot\|]$.
To B,L one can construct majorants B^*,L^* given by closed terms of HA^ω *such that if*

$$\exists n\forall x\varphi_0(x,n,\underline{a})$$

is monotone (see Definition 2.4) then it is even learnable in B^*,L^* uniformly in majorants $\underline{a}^*$ of the parameters $\underline{a}$.

Remark. Of course, instead of sequences in $[0,1]$ one can also consider sequences in any compact interval $[-C,C]$, where then the functionals B,L will additionally depend on C.

Likewise, instead of decreasing sequences we may also have increasing ones or, if the Cauchy property is changed into the existence of an approximate infimum

$$\exists n\,\forall m(a_n\le a_m+2^{-k}),$$

also arbitrary sequences in $[-C,C]$.

The next proposition shows how to convert any majorants (B^*,L^*) for a (B,L)-learnable formula into a rate of metastability. This not only guarantees a highly uniform (and for computable (B^*,L^*)) computable rate of metastability but, moreover, such a rate which has a particularly simple form (see the remark and discussion after the proposition): [11]

Proposition 2.16. *Let $\exists m^0\forall k^0\,\varphi_0(n,m,k,\underline{a})$ be a formula in the language of* HA^ω *(or* $\mathsf{HA}^\omega[X,\|\cdot\|]$*) with φ_0 being quantifier-free that is (B,L)-learnable in the sense of Definition 2.9 (uniformly in n and $\underline{a}$) and let B^*,L^* be majorants of B,L.*
Then a rate of metastability Ω (valid in $\mathcal{S}^{\omega,X}$)

$$\forall n^0\,\forall g^1\,\exists m\le_0\Omega(g,n)\;\varphi_0(n,m,g(m),\underline{a}) \qquad \text{(metastable)}$$

for[12]

$$\forall n^0\exists m^0\forall k^0\;\varphi_0(n,m,k,\underline{a}) \qquad (\varphi)$$

is given by $\Omega(B^*,L^*,\underline{a}^*)$ (uniformly in majorants $\underline{a}^*$ of the parameters $\underline{a}$), where $\tilde{g}(c):=\max\left(c,\max_{c'\le c}(g(c'))\right)$ and

$$\Omega:=\lambda B^*,L^*,\underline{a}^*,g,n\,.\,C(L^*,g,n,B^*(n,\underline{a}^*),\underline{a}^*),$$

$$C(i):=C(L^*,g,n,i,\underline{a}^*):=\begin{cases}0, & \text{if } i=0,\\ L^+(\langle\overbrace{\tilde{g}(C(i-1)+1),\ldots,\tilde{g}(C(i-1)+1)}^{i\times}\rangle,n,\underline{a}^*), & \text{else,}\end{cases}$$

[11] Note that in our examples, $\underline{a}$ will be data related to an abstract normed of Hilbert space for which (in contrast to $\underline{a}^*$) computability is not even defined.

[12] Note that in order to talk about metastability, we need one of the parameters to have type 0 and we treat it separately.

with $L^+(x) := \max\{L^*(x), x\}$.
Note that Ω *is defined using only recursion* R_0 *of type* 0 *and hence is primitive recursive in the usual sense of Kleene.*

Proof. We reason in $\mathcal{S}^{\omega,X}$. Since φ_0 is quantifier-free, there is a closed term f with $f(n,m,k,\underline{a}) = 0 \leftrightarrow \varphi_0(n,m,k,\underline{a})$. Hence, we have that (for $\underline{a}^*$ majorizing $\underline{a}$) and for the succession c_i from Definition 2.9

$$\forall n\, \exists i \leq B^*(n,\underline{a}^*)\, \forall k\ \left(f(n,c_i,k,\underline{a}) = 0\right),$$

by the assumptions of the proposition, and we need to show that

$$\forall g,n\, \exists m \leq \Omega(B^*,L^*,\underline{a}^*,g,n)\ \left(f(n,m,g(m),\underline{a}) = 0\right).$$

Now, fix any g, n and assume towards contradiction that $\Omega(B^*,L^*,\underline{a}^*,g,n)$ is not a rate of metastability, i.e. that

$$\forall m \leq \Omega(B^*,L^*,\underline{a}^*,g,n)\ \left(f(n,m,g(m),\underline{a}) \neq 0\right). \tag{3}$$

By induction on i we obtain that

$$\forall i \leq B^*(n,\underline{a}^*)\ \left(c_i \leq C(L^*,g,n,i,\underline{a}^*)\right). \tag{4}$$

The case $i = 0$ is trivial as $c_0 = C(L^*,g,n,0,\underline{a}^*) = 0$. Next, suppose that for some $1 \leq i \leq B^*(n,\underline{a}^*)$ the following holds

$$\forall j < i\ \left(c_j \leq C(L^*,g,n,j,\underline{a}^*)\right). \tag{5}$$

Denoting the smallest k s.t. $f(n,m,k,\underline{a}) \neq 0$ by x_m (if this does not exist, we have $c_i = c_{i-1}$ and are done), we obtain by (5) that[13]

$$c_i \leq L^*(\langle x_{c_0},\ldots,x_{c_{i-1}}\rangle,n,\underline{a}^*) \leq L^*(\langle \tilde{g}(c_0),\ldots,\tilde{g}(c_{i-1})\rangle,n,\underline{a}^*) \leq C(L^*,g,n,i,\underline{a}^*),$$

since by (3) we have (using that $C(i)$ is nondecreasing in i) that

$$\forall i \leq B^*(n,\underline{a}^*)\ \left(m \leq C(L^*,g,n,i,\underline{a}^*) \to f(n,m,g(m),\underline{a}) \neq 0\right)$$

and so, in particular, that

$$\forall i \leq B^*(n,\underline{a}^*)\ \left(m \leq C(L^*,g,n,i,\underline{a}^*) \to x_m \leq g(m) \leq \tilde{g}(m)\right).$$

Finally, we can infer from (4) that

$$\forall i \leq B^*(n,\underline{a}^*)\ \left(c_i \leq \Omega(B^*,L^*,\underline{a}^*,g,n)\right),$$

and therefore and by (3) also

$$\forall i \leq B^*(n,\underline{a}^*)\ \neg\forall k^0\left(f(n,c_i,k,\underline{a}) = 0\right)$$

for all majorizable $\underline{a}$ contradicting the (B,L)-learnability (uniformly in n and $\underline{a}$) of $\exists m \forall k\, \varphi_0(n,m,k,\underline{a})$ and the fact that B^* majorizes B (so that $B^*(n,\underline{a}^*) \geq B(n,\underline{a})$). □

[13] Here we simply assume that our encoding is monotone in its components. If for some reason it was not, we could use a L' which returns the maximal value among all codes coordinatewise bounded by the elements of the encoded input of L^*.

Remark 2.17. Note that Ω has essentially the following form[14]

$$(L_{n,\underline{a}^*} \circ \tilde{g})^{B^*(n,\underline{a}^*)}(0),$$
$$L_{n,\underline{a}^*} := \lambda x \,.\, L^*(x,n,\underline{a}^*).$$

Moreover, if we have such a rate of metastability for some Cauchy statement φ as considered in Remark 2.12 so that φ is monotone and a counterexample x is always greater than the witness candidate, and given any $n,\underline{a}^*$ we have an f^1 and a b^0 such that for all $\underline{a}$ that are majorized by $\underline{a}^*$

$$\forall g \exists m \leq (f \circ \tilde{g})^b(0) \varphi_0(n,m,g(m),\underline{a}), \tag{6}$$

then φ is B,L-learnable (uniformly in n and majorants $\underline{a}^*$ for $\underline{a}$) with

$$B(n,\underline{a}^*) := b, \quad L(x,n,\underline{a}^*) := f(x).$$

To prove this fact, we argue as follows. Fix arbitrary $n,\underline{a}^*$ and consider corresponding f and b. Let

$$g(m) = \min\{x \,:\, \neg\varphi_0(n,m,x,\underline{a})\},$$

if such an x exists and 0 otherwise. Note that due to the monotonicity of φ we have $g(m) \neq 0 \to \tilde{g}(m) = g(m)$. This implies that as long as there is a (the smallest) counterexample x_i to c_i, it holds that

$$c_{i+1} = L(x_i,n,\underline{a}^*) = f(x_i) = (f \circ \tilde{g})c_i = (f \circ \tilde{g})^{i+1}(0). \tag{7}$$

Given all this, assume towards contradiction that

$$\forall i \leq B(n,\underline{a}^*) \exists x \neg\varphi_0(n,c_i,x,\underline{a}), \tag{8}$$

and consider any $m \leq (f \circ \tilde{g})^b(0)$. Due to (7) and (8), we get that $m \leq c_b$ and due to the monotonicity of φ this means that there is a counterexample to m (since there is one for c_b by (8), as $B(n,\underline{a}^*) = b$), which means that $\neg\varphi_0(n,m,g(m),\underline{a})$ by definition of g. This is a contradiction to (6).

Discussion: What the main results in this section (Theorem 2.11 together with Proposition 2.16) show is that if the proof of a (monotone) Π^0_3-statement (e.g. a Cauchy statement) uses only a bounded (in the parameters) number of unnested Σ^0_1-LEM$^-$-instances but may use induction of unrestricted complexity, then we get a rate of metastability which has a remarkably simple structure (as a functional in the counterfunction g, namely only a single use of its iteration). Note that e.g. HA$^\omega$ does, of course, prove $\forall g \exists x\, \psi_0(g,x)$-sentences, and that every type-2 functional (in g) is definable in Gödel's calculus T arrives in this way. For a more detailed discussion with examples see [83].

2.2 Cauchy statements and unrestricted use of Σ^0_1-LEM

In the following, we refer to Friedman's so-called A-translation from [28] (see e.g. [71]). Since we work in the context of weakly extensional systems and the quantifier-free rule of extensionality QF-ER is not sound under the A-translation we simply add for the

[14]Note that the additional dependency on the number i of iterates in Proposition 2.16 via the length of the sequence $\langle \ldots \rangle$ can also be covered by this normal form, since – by $\tilde{g}(C(i-1)+1) \geq \tilde{g}(i) \geq i$ – the length of the sequence $\langle \ldots \rangle$ can be majorized by $\tilde{g}(C(i-1)+1)$ itself.

reminder of this section all $\mathcal{S}^\omega$-true (resp. $\mathcal{S}^{\omega,X}$-true) purely universal sentences $\mathcal{P}$ in the language of the respective system as axioms (making the use of QF-ER in proofs superfluous as it only proves universal consequences). This, anyhow, is a common device in proof mining as universal axioms do not contribute to the computational content of a proof (this has been stressed by G. Kreisel since the 50's). We denote the extension of the theory $\mathcal{T}$ by the axioms $\mathcal{P}$ by $\mathcal{T}_*$.

Lemma 2.18. *Friedman's A-translation is sound also for* $\mathsf{HA}^\omega_* + \Sigma^0_1$-LEM. *Similarly for* $\mathrm{HA}^\omega_*[X, \|\cdot\|]$ *(and related extensions) instead of* HA^ω_*.

Proof. Consider the following instance of Σ^0_1-LEM

$$\forall y \varphi_0(\underline{a}, y) \vee \exists y \neg \varphi_0(\underline{a}, y).$$

W.l.o.g assume φ_0 is atomic. It suffices to extend Friedman's proof by showing that

$$\mathsf{HA}^\omega_* + \Sigma^0_1\text{-LEM} \vdash (\Sigma^0_1\text{-LEM})^A.$$

This means we need to prove

$$\forall y(\varphi_0(\underline{a}, y) \vee A) \vee \exists y((\varphi_0(\underline{a}, y) \vee A) \to A), \tag{1}$$

in $\mathsf{HA}^\omega_* + \Sigma^0_1$-LEM. Suppose that

1. $\forall y \varphi_0(\underline{a}, y)$ holds. Then also $\forall y(\varphi_0(\underline{a}, y) \vee A)$ holds and therefore also (1).
2. $\exists y \neg \varphi_0(\underline{a}, y)$ holds. Then fix such a y. For this y we get

$$(\varphi_0(\underline{a}, y) \vee A) \to A$$

and so $\exists y((\varphi_0(\underline{a}, y) \vee A) \to A)$ holds and therefore also (1).

For $\mathrm{HA}^\omega_*[X, \|\cdot\|]$ one just has to observe that still every quantifier-free formula can be written as an atomic formula of the form $t\underline{a} =_0 0$ and that the additional axioms are all purely universal and so easily imply their own A-translation. □

For HA instead of HA^ω_* and $\mathrm{HA}^\omega_*[X, \|\cdot\|]$ (also for Σ^0_{n+1}-LEM and Σ^0_{n+2}-DNE), the next proposition is stated (without proof) in [44].

Proposition 2.19. *The theory* $\mathsf{HA}^\omega_* + \Sigma^0_1$-LEM *is closed under the* Σ^0_2-DNE *rule. Similarly for* $\mathrm{HA}^\omega_*[X, \|\cdot\|]$.

Proof. Suppose

$$\mathsf{HA}^\omega_* + \Sigma^0_1\text{-LEM} \vdash \neg\neg\exists x \forall y\, \varphi_0(\underline{a}, x, y),$$

where φ_0 is quantifier free and contains only $\underline{a}$ as free variables (in addition to x, y). Moreover, w.l.o.g we assume that φ_0 is atomic.
Rewriting the negations in terms of "$\to$" and "$\bot$" we obtain that

$$\mathsf{HA}^\omega_* + \Sigma^0_1\text{-LEM} \vdash (\exists x \forall y\, \varphi_0(\underline{a}, x, y) \to \bot) \to \bot,$$

and using Friedman's A-translation (with Lemma 2.18) that

$$\mathsf{HA}^\omega_* + \Sigma^0_1\text{-LEM} \vdash \Big(\exists x \forall y\, (\varphi_0(\underline{a}, x, y) \vee A) \to A\Big) \to A,$$

for any formula A (not containing x, y free). By setting

$$A :\equiv \exists x' \forall y' \varphi_0(\underline{a}, x', y'),$$

(we consider only this A throughout the remainder of the proof) we obtain that $\mathsf{HA}^\omega_* + \Sigma^0_1$-LEM proves

$$(\exists x \forall y\, (\varphi_0(\underline{a}, x, y) \vee \exists x' \forall y' \varphi_0(\underline{a}, x', y')) \to \exists x' \forall y' \varphi_0(\underline{a}, x', y')) \to \exists x' \forall y' \varphi_0(\underline{a}, x', y'). \tag{1}$$

Now the claim follows from

$$\Sigma^0_1\text{-LEM} \vdash \forall y (\varphi_0(\underline{a}, x, y) \vee \exists x' \forall y' \varphi_0(\underline{a}, x', y')) \to (\forall y\, \varphi_0(\underline{a}, x, y) \vee \exists x' \forall y' \varphi_0(\underline{a}, x', y')), \tag{2}$$

since using (2) the statement (1) is equivalent to

$$((\exists x \forall y \varphi_0(\underline{a}, x, y) \vee \exists x' \forall y' \varphi_0(\underline{a}, x', y')) \to \exists x' \forall y' \varphi_0(\underline{a}, x', y')) \to \exists x' \forall y' \varphi_0(\underline{a}, x', y'),$$

which is equivalent to $\exists x \forall y \varphi_0(\underline{a}, x, y)$.
To show (2) consider the following instance of Σ^0_1-LEM

$$\forall y\, \varphi_0(\underline{a}, x, y) \vee \exists y\, \neg \varphi_0(\underline{a}, x, y).$$

Now suppose that

1. $\forall y\, \varphi_0(\underline{a}, x, y)$ holds, then (2) is trivially true.
2. $\exists y\, \neg \varphi_0(\underline{a}, x, y)$ holds, then for such a y we have

   $$(\varphi_0(\underline{a}, x, y) \vee \exists x' \forall y' \varphi_0(\underline{a}, x', y')) \to \exists x' \forall y' \varphi_0(\underline{a}, x', y')$$

   and so certainly we have also that

   $$\forall y (\varphi_0(\underline{a}, x, y) \vee \exists x' \forall y' \varphi_0(\underline{a}, x', y')) \to \exists x' \forall y' \varphi_0(\underline{a}, x', y').$$

   Finally $(\forall y\, \varphi_0(\underline{a}, x, y) \vee \exists x' \forall y' \varphi_0(\underline{a}, x', y'))$ follows from $\exists x' \forall y' \varphi_0(\underline{a}, x', y')$ so (2) holds as well.

□

It is known, that a Cauchy rate is limit computable (which corresponds to Σ^0_2-DNE which – as mentioned in the introduction – is strictly stronger than Σ^0_1-LEM). However, for every provable Cauchy sequence we have that Σ^0_1-LEM is sufficient:

Proposition 2.20. *If a sequence of real numbers (a_n) (or in some PA^ω_*-definable Polish space) defined by a term of PA^ω_*, can be proved to be Cauchy in PA^ω, then the proof can be carried out already in $\mathsf{HA}^\omega_* + \Sigma^0_1$-LEM. Similarly for $\mathsf{PA}^\omega_*[X, \|\cdot\|]$ and sequences in X.*

Proof. Consider a sequence $x_{(\cdot)}$ and suppose

$$\mathsf{PA}^\omega_* \vdash \forall k \exists n \forall i, j > n\, (|x_i - x_j| \le 2^{-k}).$$

Then by the Kuroda negative translation (see e.g. [71]) we obtain that

$$\mathsf{HA}^\omega_* \vdash \forall k \neg\neg \exists n \forall i, j > n\, (|x_i - x_j| \le 2^{-k}).$$

By Proposition 2.19 this implies that

$$\mathsf{HA}^\omega_* + \Sigma^0_1\text{-LEM} \vdash \forall k \exists n \forall i, j > n\, (|x_i - x_j| \le 2^{-k})$$

(recall that $\le_{\mathbb{R}} \in \Pi^0_1$). □

2.3 Which Cauchy statements are effectively learnable and which are not

Proposition 2.21 (Implications between different bounding information for Cauchy statements). *Let (x_n) be a Cauchy sequence in a metric space (X,d).*

1. *A rate of convergence is a bound on the number of fluctuations.*
2. *A bound for the number of fluctuations is a bound B on the number of mind changes to learn a rate of convergence (with a simple projection function as learning procedure L).*
3. *Primitive recursively (in the ordinary sense of Kleene) in majorants B^*, L^* of functionals B, L such that the Cauchy rate is (B, L)-learnable one can obtain a rate of metastability.*

Proof. Consider a Cauchy sequence $x_{(\cdot)}$.

1. Let b be a rate of convergence, i.e.

$$\forall k \forall n, m \geq b(k) \left(d(x_n, x_m) \leq 2^{-k}\right).$$

   Then $b(k)$ is also a bound on the number of 2^{-k} fluctuations, since any fluctuation has to occur before $b(k)$ (i.e. that one of the indexes of the fluctuation has to be smaller than $b(k)$) and there can be at most $b(k)$ many fluctuations indexed within $[0; b(k)]$.

2. Let b be a bound on the number of 2^{-k} fluctuations, i.e.

$$\forall k \forall n > b(k) \forall i, j \neg \mathrm{Fluc}_{2^{-k}}(n, i, j).$$

   Then $b(k)$ is also a bound on the number of mind changes to learn a rate of convergence, since for $L(n) := n + 1$ we have that

$$\forall k\ \exists l \leq b(k)\ \forall n, m > c_l\ \left(d(x_n, x_m) \leq 2^{-k}\right). \tag{BE}$$

   Formally, $L(i, x_{(\cdot)}, k) := i$, and (where – again – to have φ_0 quantifier-free we officially have to use the 2^{-k-1}-rational approximation $\widehat{d(x_n, x_m)}(k+1)$ of $d(x_n, x_m)$)

$$\varphi_0(j(n,m), c_i, x_{(\cdot)}, k) :\equiv \left((n > c_i \wedge m > c_i) \rightarrow d(x_n, x_m) \leq 2^{-k}\right),$$

   where $j(n, m)$ is the Cantor pairing function. The statement (BE) can be inferred from the fact that each mind change (c_i) corresponds to a (different) fluctuation (as it is based on a counterexample for $d(x_n, x_m) \leq 2^{-k}$, whose both indexes are greater than the last c_i).

3. Follows directly from Proposition 2.16.

□

In the rest of this section we show that the hierarchy in Proposition 2.21 between the four different quantitative notations for Cauchy sequences discussed in the introduction is strict. That an effective bound on the number of fluctuations does not imply an effective rate of convergence, follows already from the existence of Specker sequences. We can also use the following very simple example with a 2^{-k}-fluctuation bound k and no effective rate of convergence, since such a rate would decide the halting problem.

Proposition 2.22 ($\alpha_{(\cdot)}$). *We take the Cantor pairing function j and set*

$$\alpha_{j(k,n)} := \begin{cases} 2^{-k}, & \text{if} T(k,0,n), \\ 0, & \text{else,} \end{cases}$$

where T is the primitive recursive Kleene T-predicate. Then (α_n) is a convergent (towards 0) primitive recursive sequence of rationals in $[0,1]$ with 2^{-k}-fluctuation bound k which has no computable rate of convergence.

We next construct primitive recursive sequence $\beta_{(\cdot)}$ of rational numbers in $[0,1]$ with an effectively (even primitive recursively) learnable Cauchy rate (so in particular with a primitive recursive rate of metastability), which has no computable bound on fluctuations (this example is not captured by the rough sketch of Avigad and Rute as here the number of the oscillations is determined by the length of the computation, not by the index of the machine as suggested in [10]). Furthermore, we also give an example of a primitive recursive (in the ordinary sense) sequence $\gamma_{(\cdot)}$ of rational numbers in $[0,1]$ which (provably in the fragment of PA based on Σ^0_1-IA only) converges to 0 (and so has a primitive recursive in the sense of Kleene rate of metastability for the convergence towards 0) which does not have an effectively learnable Cauchy rate.

An effectively learnable sequence with no computable bound on fluctuations

Definition 2.23 ($\beta_{(\cdot)}$). We fix a primitive recursive surjective encoding of triples which is monotone in the third component satisfying $\langle k,n,m\rangle \geq k,n,m$ and set

$$\beta_{\langle k,n,l\rangle} := \begin{cases} 2^{-k}, & \text{if} T(k,0,n) \wedge l \leq n \wedge l \text{ is even,} \\ 0, & \text{else.} \end{cases}$$

In the next propositions we will show that the sequence $\beta_{(\cdot)}$

- is Cauchy (in fact, it converges to zero) – Proposition 2.24,
- its Cauchy rate is effectively learnable – Proposition 2.27,
- there is no computable (in ϵ and β) bound on the number of ϵ-fluctuations – Proposition 2.31.

Proposition 2.24. *The sequence $\beta_{(\cdot)}$ is convergent towards 0, provably in $\mathsf{HA}^\omega + \Sigma^0_1$-LEM$^-$. More precisely we show that*

$$\mathsf{HA}^\omega \vdash \forall k\Big(\forall m \leq k\,(\exists u\, T(m,0,u) \vee \forall v\, \neg T(m,0,v)) \to \exists n \forall x \geq n\big(\beta_x \leq 2^{-k}\big)\Big).$$

Proof. Consider the terminating computations on input 0 of the Turing machines encoded by $0,\ldots,k$. Then for every k there is an n corresponding to the code of the longest such computation. W.l.o.g. we can assume that $n \geq k$ (otherwise set $n := k$). This means we have that

$$n \geq k \wedge \forall n' \forall k' \leq k\,\big(T(k',0,n') \to n' \leq n\big). \tag{9}$$

Now, set

$$c(k) := \max\{\langle k',n',l'\rangle \,:\, n' \leq n,\, k' \leq k,\, l' \leq n'\}.$$

Then c is even a rate of convergence, since

$$\langle k',n',l'\rangle > c(k) \rightarrow k' > k \vee (k' \leq k \wedge n' > n) \vee (k' \leq k \wedge n' \leq n \wedge l' > n')$$
$$\rightarrow k' > k \vee \beta_{\langle k',n',l'\rangle} = 0 \rightarrow \beta_{\langle k',n',l'\rangle} < 2^{-k}.$$

These arguments are constructive, except for the existence of the longest computation n. This existence is a consequence of Σ^0_1-LEM$^-$ and Π^0_1-CP$^-$, where Π^0_1-CP is the bounded collection principle for Π^0_1-formulas (also called $B\Sigma^0_2$ in the literature) which is easily provable by induction in HA$^\omega$. Consider the following $k+1$ instances of Σ^0_1-LEM$^-$:

$$\forall j \leq k\ (\exists n T(j,0,n) \vee \forall m \neg T(j,0,m))$$

which over HA$^\omega$ implies

$$\forall j \leq k \exists n_j \forall m\ (T(j,0,n_j) \vee \neg T(j,0,m)).$$

By an application of Π^0_1-CP$^-$, this in turn implies

$$\exists n \forall j \leq k (\exists n' \leq n T(j,0,n') \vee \forall m \neg T(j,0,m)),$$

(consider $n = \max\{n_j : j \leq k\}$).
This shows that the convergence is provable in HA$^\omega$+Σ^0_1-LEM$^-$ and the convergence up to an error 2^{-k} in HA$^\omega$ uses only $k+1$ instances of Σ^0_1-LEM$^-$. □

Lemma 2.25. *For a quantifier-free formula φ_0 with parameters only of type 0, we have that*

$$\mathrm{G_3A}^\omega + \Sigma^0_1\text{-IA}^- \vdash \forall x^0\ \exists u^0 \forall \tilde{x} \leq x (\forall y^0 \varphi_0(\tilde{x},y) \vee \exists \tilde{u} \leq u \neg \varphi_0(\tilde{x},\tilde{u})).$$

where G$_3$A$^\omega$ is the finite type extension of Kalmar-elementary arithmetic (based on quantifier-free induction only but with classical logic) from [63] *(see also* [71]*).*

Proof. See [71] Lemma 3.18. □

Remark 2.26. Σ^0_1-IA$^-$ is (over G$_3$A$^\omega$) strictly weaker than Π^0_1-CP$^-$ but the proof in Lemma 2.25 needs Σ^0_2-DNE and so more of classical logic than necessary in the proof based on Π^0_1-CP$^-$. In general, it seems that considering the extraction of computational content from proofs, often some amount of classical logic can be reduced on the cost of more recursion.
If one is interested (only) in a classical proof, we obtain due to Lemma 2.25 (simply consider $\varphi_0(x,y) :\equiv \neg T(x,0,y)$) a proof of the convergence of $\beta_{(\cdot)}$ without the use of Π^0_1-CP, which can be formalized in G$_3$A$^\omega$ + Σ^0_1-IA$^-$.

Proposition 2.27. *The rate of convergence is effectively learnable in k, i.e. there are total (elementary) recursive functions B and L, s.t. for any k we have that*

$$\forall k\ \exists n \leq B(k)\ \forall m > c_n \quad (\beta_m \leq 2^{-k}),$$

where $c_{(\cdot)}$ is defined as in Definition 2.4 with

$$\varphi_0(x,n,k) :\equiv\ x > n \rightarrow \beta_x \leq 2^{-k}.$$

Proof. Obviously, this follows already from Proposition 2.24. Also, it is easy to see that the rate is (B, L)-learnable with the following B and L:

$$B(k) := k + 1$$
$$L(n, k) := \langle k, n, n\rangle + 1.$$

Let $x \geq L(n,k) > \langle k, n, n\rangle$ be a counterexample. Then (using the definition of $\beta_{(\cdot)}$)

$$j_1(x) \leq k \wedge (j_2(x) > n \vee j_3(x) > n) \wedge j_3(x) \leq j_2(x)$$

and so

$$j_1(x) \leq k \wedge j_2(x) > n \wedge j_3(x) \leq j_2(x).$$

The 2nd conjunct implies $j_2(x) > j_2(n)$ and hence $j_1(x) \neq j_1(n)$ if n is a preceding counterexample. However, for numbers $\leq k$ this can happen at most k-many times. Hence $B(k) := k + 1$ and L do the job. □

Corollary 2.28. *$\beta_{(\cdot)}$ has a primitive recursive (in the ordinary sense of Kleene) rate of metastability for the convergence towards 0.*

Proof. One can apply Proposition 2.16 to convert the bounds (B, L) from Proposition 2.27 (which are trivially self-majorizing using standard monotonicity properties of the triple coding) into a primitive recursive rate of metastability. Alternatively, one can use that by Lemma 2.25 the convergence of $\beta_{(\cdot)}$ towards 0 is provably in $G_3A^\omega + \Sigma^0_1\text{-IA}^-$ and so a fortiori in $\widehat{\text{PA}}^\omega\restriction$+QF-AC (see [71], Prop.3.31). Then proposition 10.54 in [71] implies the extractability of a primitive recursive rate of metastability (the latter being essentially the rate of metastability for the statement in Lemma 2.25 which is computed in [71][Prop.3.19]). □

Remark 2.29. One can obtain such a rate of metastability for $\beta_{(\cdot)}$ directly, using previous results of U. Kohlenbach. Define $\tilde{f}(n) := \max\{f(n), n\}$ and $f_k(n) := \tilde{f}(\langle k, n, n\rangle + 1)$ and let

$$\Psi(k, f) := \langle k, \Phi^* k f_k, \Phi^* k f_k\rangle + 1.$$

Then Ψ is a rate of metastability for the convergence of $\beta_{(\cdot)}$ towards 0, i.e.:

$$\forall k, f\, \exists n \leq \Psi(k, f)\, \forall z \in [n, \tilde{f}(n)]\ \ (|\beta_z| < 2^{-k}).$$

For details see [83].

Lemma 2.30 (Termination causes at least n fluctuations). *Suppose the kth-machine terminates on 0 with computation encoded by n (i.e. $T(k, 0, n)$ holds). Then the sequence $\beta_{(\cdot)}$ contains at least n many 2^{-k}-fluctuations.*

Proof. Consider the tuples of indexes $\underline{i}$, $\underline{j}$, s.t. $i_l := \langle k, n, l\rangle$, $j_l := \langle k, n, l+1\rangle$ and $l + 1 \leq n$. Then we have by definition of $\bar{\beta}_{(\cdot)}$ (using the monotonicity of the encoding in l) that $\text{Fluc}_{\beta_{(\cdot)}, 2^{-k}}(n, \langle\underline{i}\rangle, \langle\underline{j}\rangle)$. □

Proposition 2.31. *There is no computable bound on the fluctuations of $\beta_{(\cdot)}$.*

Proof. Suppose b_k is a bound on the number of fluctuations by 2^{-k}, then b_k can be used to effectively compute whether the kth Turing machine terminates on input 0 as follows. Let the machine run until the code of the computation reaches b_k (or until it stops). If it terminated, we are done.
Now suppose it terminates with a computation encoded by some $n > b_k$. Then $\beta_{(\cdot)}$ would have at least n many 2^{-k}-fluctuations by Lemma 2.30, which is a contradiction. Therefore if the machine does not terminate with a code of computation at most b_k it does not terminate at all. □

Metastability does not imply effective learnability

In the next propositions we define a primitive recursive sequence $\gamma_{(\cdot)}$ of rational numbers in $[0, 1]$ (defined in Corollary 2.40 using Definition 2.36) that

- is Cauchy (in fact, it converges to zero) – by Proposition 2.37,
- has a primitive recursive rate of metastability of its convergence towards 0 – by Proposition 2.37,
- has no effectively learnable Cauchy rate – by Corollary 2.40.

We use the upper index as a name extension (like $\underline{k}^K$, meaning a tuple $\underline{k}$ corresponding to a particular K) and as iteration of functions (like $f^n(x)$, meaning we iterate the function f n-many times with the starting point x). When unclear, we use the notation $(f)^n$ to make explicit, that we mean the iteration.

Definition 2.32. 1. For any function $f : \mathbb{N}^2 \to \mathbb{N}$ we define

$$\widehat{f}(k,n) := 0, \text{ if } f(k,n) = 0 \wedge \forall m < n\, (f(k,m) \neq 0) \text{ and } \widehat{f}(k,n) := 1, \text{ otherwise.}$$

Note that $\widehat{f}(k,\cdot)$ has a root iff $f(k,\cdot)$ has one but also that it has at most one root.

2. For any f as above and $x, y \in \mathbb{N}$ define

$$y_{f,x} := \begin{cases} \max\{y' \leq y : \exists x' \leq x\, (y' = \min\{y'' \leq y \,:\, f(x',y'') = 0\,\})\}, & \text{if such } y' \text{ exists} \\ x, & \text{else.} \end{cases}$$

Then for $p = j(y, u)$ define

$$\varphi_1(f,x,p,z) :\equiv \forall \tilde{x} \leq x \exists \tilde{y} \leq y \forall \tilde{z} \leq z \big(f(\tilde{x},\tilde{y}) = 0 \vee f(\tilde{x},\tilde{z}) \neq 0\big)$$

and

$$\varphi_2(f,x,p,z) :\equiv \forall \tilde{y} \leq y_{f,x} \exists \tilde{u} \leq u \forall \tilde{z} \leq z \big(f(\tilde{y},\tilde{u}) = 0 \vee f(\tilde{y},\tilde{z}) \neq 0\big)$$

and, finally,[15]

$$\varphi_0 :\equiv \varphi_1 \wedge \varphi_2, \quad \varphi :\equiv \forall f \leq_1 \forall x^0 \exists p \forall z \varphi_0(\widehat{f},x,p,z).$$

[15] In connection with f we write the type 1 even though it officially is the type $0 \to (0 \to 0)$.

Note that the φ_1-part of φ combines a sequence of instances of Σ^0_1-LEM with induction (in the form of Π^0_1-CP) and that the φ_2-part repeats this construction but taking (essentially) the result 'y' from φ_1 as input thereby making it no longer possible to give a computable bound on the number of instances of Σ^0_1-LEM used in total. We will show in the next proposition that φ is not (B,L)-learnable by showing that it implies over a system as weak as G_3A^ω (which does not allow for the iteration of a function variable g)

$$\forall g^1 \exists y^0 \, (y = g^{g^x(0)}(0))$$

which grows too fast as a functional in g to be derivable from a rate of metastability for φ having the simple form from Remark 2.17 whose existence would follow from the (B,L)-learnability of φ. Here it is crucially used that computable functionals B,L in the function parameter f which can be taken to be bounded by 1 can be effectively majorized by bounds which no longer depend on f.

Proposition 2.33. *φ is provable using Σ^0_1-LEM$^-$ combined with Π^0_1-CP$^-$ (uniformly in f treated as parameter) but is not effectively learnable.*

Proof. We first show that φ is provable: by a suitable instance Σ^0_1-LEM$(t(f))$ of Σ^0_1-LEM$^-$ one obtains

$$\forall x \forall \tilde{x} \le x \exists y \forall z \big(\exists \tilde{y} \le y \, \widehat{f}(\tilde{x},\tilde{y}) = 0 \vee \forall \tilde{z} \le z \, \widehat{f}(\tilde{x},\tilde{z}) \neq 0\big)$$

and so by a suitable instance Π^0_1-CP$(s(f))$

$$\forall x \exists y \forall \tilde{x} \le x \forall z \big(\exists \tilde{y} \le y \, \widehat{f}(\tilde{x},\tilde{y}) = 0 \vee \forall \tilde{z} \le z \, \widehat{f}(\tilde{x},\tilde{z}) \neq 0\big)$$

which implies

$$\forall x \exists y \forall z \forall \tilde{x} \le x \exists \tilde{y} \le y \forall \tilde{z} \le z \, \big(\widehat{f}(\tilde{x},\tilde{y}) = 0 \vee \widehat{f}(\tilde{x},\tilde{z}) \neq 0\big).$$

Now we repeat the same argument with $y_{f,x}$ instead of x.
To show that φ is not learnable we proceed in three steps.

Step 1 We will show that (informally speaking, since formally we cannot express function iteration and the conclusion would have to use ψ_0 and ψ^y_0 defined below)

$$G_3A^\omega \vdash \forall g \forall x \, \big(\exists p \forall z \varphi_0(f_g,x,p,z) \to \exists y' (y' = g^{g^x(0)}(0))\big), \qquad \text{(GA)}$$

for

$$f_g(b,d) := \begin{cases} 0, & \text{if } \mathrm{lh}(d) = b+1 \wedge d_0 = 0 \wedge \forall i < b \, (d_{i+1} = g(d_i), \\ 1, & \text{else.} \end{cases}$$

Formally, $\exists y' (y' = g^{g^x(0)}(0))$ is to be read as

$$\exists y,u \, (f_g(x,y) = 0 \wedge f_g(y_x,u) = 0), \qquad \text{(GA*)}$$

where then $y' := u_{y_x}$. Note that we have $f_g =_1 \widehat{f_g}$. To show (GA) fix arbitrary g^1 and x^0. Now assume $\exists p \forall z \varphi_0(f_g,x,p,z)$ and let us fix such a $p = j(y,u)$ to obtain:

$$\forall z \forall \tilde{x} \le x \exists \tilde{y} \le y \forall \tilde{z} \le z \big(f_g(\tilde{x},\tilde{y}) = 0 \vee f_g(\tilde{x},\tilde{z}) \neq 0\big) \wedge \qquad (10)$$

$$\forall z \forall \tilde{y} \le y_{f_g,x} \exists \tilde{u} \le u \forall \tilde{z} \le z \big(f_g(\tilde{y},\tilde{u}) = 0 \vee f_g(\tilde{y},\tilde{z}) \neq 0\big). \qquad (11)$$

We now show by quantifier-free induction that

$$\forall x' \leq x \exists y' \leq y\, f_g(x', y') = 0. \tag{12}$$

Note that $\exists y'\, f_g(x', y') = 0$ implies that $y'_{x'} = g^{x'}(0)$. The case $x' = 0$ is trivially satisfied by $y' = \langle 0 \rangle$. Then $y' \leq y$ by (10). So suppose for some $x' < x$ we have $\exists y' \leq y\, f_g(x', y') = 0$. Then we can set

$$z := y' * \langle g(y'_{x'}) \rangle = \langle y'_0, y'_1, \ldots, g(y'_{x'}) \rangle$$

to get $f_g(x' + 1, z) = 0$ which concludes the proof of (12) since – again by (10) – $z \leq y$. Note, furthermore, that for $y' \leq y$ (and $x' \leq x$),

$$f_g(x', y') = 0 \;\rightarrow\; y' \leq y_{f,x}.$$

So we have even that

$$\forall x' \leq x \exists y' \leq y_{f_g,x}\, f_g(x', y') = 0. \tag{13}$$

Now, let y^* denote the y' which satisfies (13) for $x' = x$ and note that $y^*_x \leq y^* \leq y_{f_g,x}$. By quantifier-free induction we show that

$$\forall x' \leq y_{f_g,x} \exists u' \leq u\, f_g(x', u') = 0. \tag{14}$$

The case $x' = 0$ is again trivially satisfied by $u' = \langle 0 \rangle \leq u$ (using (11)). So suppose for some $x' < y_{f_g,x}$ that $\exists u' \leq u\, f_g(x', u') = 0$. Then we can set

$$z := u' * \langle g(u'_{x'}) \rangle = \langle u'_0, u'_1, \ldots, g(u'_{x'}) \rangle$$

to get $f_g(x' + 1, z) = 0$, which by (11) implies

$$\exists \tilde{u} \leq u\, f_g(x' + 1, \tilde{u}) = 0$$

and so concludes the proof of (14). Applying (14) we obtain (for $x' = y^*_x$)

$$\exists u' \leq u\, (f_g(x, y^*) = 0 \wedge f_g(y^*_x, u') = 0),$$

which concludes the proof of (GA*) and, therefore, also the proof of (GA).

Step 2 We investigate the terms witnessing the implication (GA). By prenexation we obtain

$$\mathrm{G_3A}^\omega \vdash \forall g, x, p \exists y', z\, (\varphi_0(f_g, x, p, z) \rightarrow y' = g^{g^x(0)}(0)),$$

and, therefore, by program extraction theorems (see Corollary 3.1.3 in [63]), we get closed terms s and t in $\mathrm{G_3A}^\omega$, s.t.

$$\forall g, x, p\, (\varphi_0(f_g, x, p, sgxp) \rightarrow tgxp = g^{g^x(0)}(0)). \tag{15}$$

Step 3 Finally, we show that with sufficiently large (in the sense of growth) g, this contradicts the effective learnability of φ. Suppose namely that φ were learnable by computable functionals $B(f,x), L(y,f,x)$. Then also

$$B^*(x) := \sup\{B(f,\tilde{x}) : f \leq_1 1, \tilde{x} \leq x\} \text{ and}$$
$$L^*_x(y) := \max\{y, \sup\{L(\tilde{y},f,\tilde{x}) : f \leq_1 1, \tilde{y} \leq y, \tilde{x} \leq x\}\}$$

are computable (in x resp. in x,y) and majorize B, L. Now by Proposition 2.16 and the fact that f_g is trivially majorized by 1 we get, in particular, that

$$\exists p \leq \Omega(B^*, L^*_x, h_x, x)\ \varphi_0(f_g, x, p, sgxp),$$

for any g and x, by setting

$$h^1_x := \lambda p \,.\, sgxp.$$

So, we obtain together with (15) that

$$\forall g, x \exists p_x \leq \Omega(B^*, L^*_x, h_x, x)\ \left(tgxp_x = g^{g^x(0)}(0)\right).$$

Since s and t are closed terms of G_3A^ω and the variables g, x, p_x have types ≤ 1 by normalization arguments (see Corollary 2.2.24 and Remark 2.2.25 in [63]) we know that there is a constant D, s.t. (for any g that majorizes $\lambda n.2^n$)

$$\forall x, v\ \left(\tilde{g}^D(x+v) \geq sgxv, tgxv\right).$$

Since we may assume that $\tilde{g}(n) > n$ and, therefore, $\tilde{g}^x(v) \geq x + v$, this yields

$$\forall x, v\ \left(\tilde{g}^{D+x}(v) \geq sgxv, tgxv\right).$$

As a consequence, we get (using the Ω-definition and that that B^*, L^*_x are selfmajorizing and that we may assume $L^*_x(y) \geq y$) that for all x

$$\tilde{g}^{D+x}\left(\Omega(B^*, L^*_x, \tilde{g}^{D+x}, x)\right) \geq \tilde{g}^{D+x}\left(\Omega(B^*, L^*_x, h_x, x)\right) \geq tgxp_x = g^{g^x(0)}(0).$$

By the definition of Ω (see also Remark 2.17)

$$\tilde{g}^{D+x}(\Omega(B^*, L^*_x, \tilde{g}^{D+x}, x) \leq \tilde{g}^{D+x}\left((L^*_x \circ \tilde{g}^{D+x})^{B^*(x)}(0)\right) \leq$$
$$\tilde{g}^{D+x}\left((L^*_x \circ \tilde{g})^{(D+x)B^*(x)}(0)\right) \leq (L^*_x \circ \tilde{g})^{\widehat{B^*}(x)}(0)$$

and so (for all x)

$$g^{g^x(0)}(0) \leq (L^*_x \circ \tilde{g})^{\widehat{B^*}(x)}(0)$$

where $\lambda x, y. L^*_x(y)$ and $\widehat{B^*}(x) := (D+x)(B^*(x)+1)$ are fixed total recursive functions that do not depend on g which is not possible for sufficiently fast growing g. □

Corollary 2.34. *Let φ_0 be as in the previous Proposition. If for a quantifier free formula ψ_0 (with no hidden parameters)*

$$G_3A^\omega + \text{QF-AC} \vdash \forall f \leq 1 \forall x^0 (\exists y^0 \forall z^0 \psi_0(\xi(f), \chi(x), y, z) \to \exists p^0 \forall z^0 \varphi_0(\widehat{f}, x, p, z)),$$

where ξ and χ are closed terms of G_3A^ω, then $\forall f \leq 1 \forall x^0 \exists y^0 \forall z^0 \psi_0(f, x, y, z)$ is also not effectively learnable. Here

$$\text{QF-AC}: \ \forall x \exists y\, F_0(x,y) \to \exists f \forall x\, F_0(x, f(x)),$$

with quantifier-free F_0 and x, y of arbitrary types.

Proof. This follows analogously from [63] (Corollary 3.1.3.) and our Proposition 2.16 as in the proof of Proposition 2.33. □

Remark 2.35. We can prove in $G_3A^\omega + \Sigma^0_1$-CP (which is included in $G_3A^\omega +$ QF-AC^{0,0}) that φ in Proposition 2.33 is actually equivalent to its monotone version,

$$\tilde{\varphi} \equiv \forall f^1 \leq 1 \forall x^0 \exists q \forall z \exists p \leq q \forall \tilde{z} \leq z \varphi_0(\widehat{f}, x, p, z).$$

Since this equivalence holds also pointwise (in f, x), we can use Corollary 2.34 to infer that there is actually a monotone formula, which is not effectively learnable.

Definition 2.36. Define (using a surjective quadruple coding) a primitive recursive sequence of rational numbers in $[0,1]$ by

$$\gamma(f)_{\langle k,n,i,m\rangle} := \begin{cases} 2^{-k}, & \text{if } \widehat{f}(k,n) = 0 \wedge i \leq n \wedge \widehat{f}(i,m) = 0, \\ 0, & \text{else.} \end{cases}$$

Proposition 2.37. *The sequence $(\gamma(f)_z)_{z\in\mathbb{N}}$ converges to 0 but the formula stating the existence of a Cauchy point for any f*

$$\psi := \forall f^1 \leq_1 1, x^0 \exists z \forall k \geq z \, (\gamma(f)_k < 2^{-x}).$$

is not effectively learnable. However, there is a primitive recursive (in the ordinary sense of Kleene) rate of metastability for the convergence of $\gamma(f)_{(\cdot)}$ towards 0, which does not depend on f.

Proof. The existence of a metastability rate follows from the fact that $\gamma(f)_{(\cdot)}$ converges to 0 for any $f \leq_1 1$. Moreover, since the proof can be formalized in $G_3A^\omega + \Sigma^0_1$-IA there is a primitive recursive rate and since f is trivially majorizable it is also clear that there is even a primitive recursive rate which does not depend on f. (In Remark 2.38 below, we actually give such a rate explicitly.)
To show the unlearnability, due to Corollary 2.34 it suffices to show that

$$G_3A^\omega + \text{QF-AC}^{0,0} \vdash \forall f \leq 1 \forall x^0 (\exists z \forall k \geq z \, (\gamma(f)_k < 2^{-x}) \to \exists p \forall z' \varphi_0(\widehat{f}, x, p, z')).$$

To prove φ, fix arbitrary f^1 and x^0 and suppose that

$$\exists z \forall k \geq z (\gamma(f)_k < 2^{-x}).$$

Moreover, assume towards contradiction

$$\exists \tilde{x} \leq x \exists a \geq \max(z, x) \widehat{f}(\tilde{x}, a) = 0. \tag{16}$$

Since $a \geq \tilde{x}, z$, this implies that $k := \langle \tilde{x}, a, \tilde{x}, a \rangle \geq z$ and $\gamma(f)_k = 2^{-\tilde{x}}$, which is a contradiction.
Hence we can conclude that

$$\forall \tilde{x} \leq x (\exists \tilde{y} < \max(x,z) \widehat{f}(\tilde{x}, \tilde{y}) = 0 \vee \forall a \widehat{f}(\tilde{x}, a) \neq 0),$$

which is equivalent to

$$\forall \tilde{x} \leq x \forall a \exists \tilde{y} < \max(x,z) \, (\widehat{f}(\tilde{x}, \tilde{y}) = 0 \vee \forall \tilde{a} \leq a \, \widehat{f}(\tilde{x}, \tilde{a}) \neq 0). \tag{17}$$

Next, set $y := \max(x,z)$ and assume towards contradiction that

$$\exists \tilde{y} \leq y_{\widehat{f},x} \exists a \geq z \widehat{f}(\tilde{y},a) = 0. \tag{18}$$

Recall that

$$y_{\widehat{f},x} := \begin{cases} \max\{y' \leq y : \exists x' \leq x\,(y' = \min\{y'' \leq y : \widehat{f}(x',y'') = 0\})\}, & \text{if such } y' \text{ exists} \\ x, & \text{else.} \end{cases} \tag{19}$$

Note that if $y_{\widehat{f},x} = x$, then φ follows already from (17). Otherwise, denote the smallest $x' \leq x$ for which $\widehat{f}(x', y_{\widehat{f},x}) = 0$ by $\tilde{x}$. Then $k := \langle \tilde{x}, y_{\widehat{f},x}, \tilde{y}, a\rangle \geq z$ and $\gamma(f)_k = 2^{-\tilde{x}}$, which is a contradiction.
Finally, $\exists p \forall z' \varphi_0(\widehat{f}, x, p, z')$ follows from not-(18) and (17) (with $y := \max(x,z)$, $p := \langle y, z\rangle$).

□

Remark 2.38. Similarly as before, we can give an explicit such rate of metastability for the convergence of $(\gamma(f)_z)$ toward 0. As in Remark 2.29, there is a $\Phi_f xg \leq \Phi^* xg := \max\{g^i(0) : i \leq x+1\}$ such that

$$\forall x, f, g \forall \tilde{x} \leq x\,(\exists y \leq \Phi_f xg\,(\widehat{f}(\tilde{x},y) = 0) \vee \forall z \leq g(\Phi_f xg)\,(\widehat{f}(\tilde{x},z) \neq 0)).$$

Define

$$\Phi_1 xg := \Phi_f\Big(x, \lambda y. g_y(\Phi_f(y, g_x))\Big),$$
$$\Phi_2 xg := \Phi_f(\Phi_1 xg, g_{\Phi_1 xg}),$$

where $g_y(n) := g(y,n)$. Then

$$\forall x, f, g\, \forall \tilde{x} \leq x \,\forall \tilde{y} \leq \Phi_1 xg \begin{cases} \Big((\exists y \leq \Phi_1 xg(\widehat{f}(\tilde{x},y) = 0) \vee \forall z \leq g(\Phi_1 xg, \Phi_2 xg)\,(\widehat{f}(\tilde{x},z) \neq 0)) \wedge \\ (\exists u \leq \Phi_2 xg\,(\widehat{f}(\tilde{y},u) = 0) \vee \forall z \leq g(\Phi_1 xg, \Phi_2 xg)\,(\widehat{f}(\tilde{y},z) \neq 0))\Big). \end{cases}$$

This implies

$$\forall z(\Phi_1 xg < z \leq g(\Phi_1 xg, \Phi_2 xg) \to \forall \tilde{x} \leq x\,(\widehat{f}(\tilde{x},z) \neq 0))$$

and

$$\forall z(\Phi_2 xg < z \leq g(\Phi_1 xg, \Phi_2 xg) \to \forall \tilde{x} \leq \Phi_1 xg\,(\widehat{f}(\tilde{x},z) \neq 0)).$$

Define $\tilde{g}(n) := \max\{g(n), n\}$ and $g_k(n,m) := \tilde{g}(\langle k,n,n,m\rangle + 1)$ and

$$\Psi(k,g) := \langle k, \Phi_1^* k g_k, \Phi_1^* k g_k, \Phi_2^* k g_k\rangle + 1,$$

where Φ_i^* is defined as Φ_i but with Φ^* and $(g_k)^M$ instead of Φ_f and g_k. Then

$$\forall k, g, f\, \exists n \leq \Psi(k,g)\, \forall z \in [n, \tilde{g}(n)]\ \ (|\gamma(f)_z| < 2^{-k}),$$

as shown in [83].
Note that the 2-nested use of primitive recursive iteration hidden in the 2-nested application of Φ_f in the definition of Φ_i very much resembles the basic structure of the rate of metastability extracted from a concrete proof in ergodic theory in next chapter (see also the discussion in the introduction in [107]).

Corollary 2.39. *Let $\gamma(e)_{(\cdot)}$ be defined as $\gamma(f)_{(\cdot)}$ but with $\widehat{f}(x,y)=0$ being replaced by*

$$P(e,x,y) :\equiv \mathrm{lh}(y) = x+1 \wedge y_0 = 0 \wedge \forall i < x(T(e,y_i,y_{i+1})).$$

Then $(\gamma(e)_n)$ is a sequence of rational numbers in $[0,1]$ that converges to 0 but the formula stating the existence of a Cauchy point of $\gamma(e)_{(\cdot)}$ for any number e

$$\psi := \forall e^0, x^0 \exists z \forall k \geq z\ (\gamma(e)_k < 2^{-x}).$$

is not effectively learnable.

Proof. First note that the convergence (towards 0) of $\gamma(e)_n$ follows from this property of $\gamma(f)_n$ since taking $f(x,y) := 0$, if $P(e,x,y)$, and $f(x,y) := 1$, otherwise, both sequences coincide (note that $f = \widehat{f}$).
Let e by a code of a total recursive function and define $g(x) := \mu y\,.\,T(e,x,y)$. The arguments of both Proposition 2.33 and Proposition 2.37 then remain valid, except the fact that the set of Gödel numbers e – in contrast to f_g – is not majorizable. We also need the additional assumption $\forall x\,(T(e,x,g(x)))$ in (GA), which expresses that $g(x) := \mu y\,.\,T(e,x,y)]$ defines a total function. This assumption, however, does not contribute in the course of the functional interpretation argument applied to (GA) ('Step 2' in the proof of Proposition 2.33) as it is purely universal. So as before, the learnability would lead to a constant D and total recursive functions $\lambda e,x,y.L^*_{e,x}(y)$, B^* such that

$$\forall e \forall g \left(\forall x^0\, T(e,x,g(x)) \to \forall x^0\ (L^*_{e,x} \circ \tilde{g})^{\widehat{B^*}(e,x)}(0) \geq g^{g^x(0)}(0))\right)$$

where $\widehat{B^*}(e,x) := (D+x+e)(B^*(x)+1)$.
We can argue similarly as in the proof of Proposition 2.33, since for any fixed g given by some e as above, eventually we have $x > e$, so we can simply choose a total recursive g which grows much faster than $L^*_{x,x}(x)$ and $B^*(x,x)$. □

Corollary 2.40. *Define the primitive recursive sequence of rational numbers in $[0,1]$ (using the Cantor pairing function)*

$$\gamma_{j(e,n)} := \gamma(e)_n \cdot 2^{-e}.$$

This sequence converges to 0 (provably in $G_3A^\omega + \Sigma^0_1$-IA^- and hence with a primitive recursive – in the sense of Kleene – rate of metastability) but the formula stating the existence of a Cauchy point of $\gamma_{(\cdot)}$

$$\psi := \forall x^0 \exists z \forall k \geq z\ (\gamma_k < 2^{-x}).$$

is not effectively learnable.

Proof. Let $\rho_e(x)$ be a rate of convergence for $(\gamma(e)_n)_n$ and define $\rho(x) := \max\{\rho_{\tilde{x}}(x) : \tilde{x} \leq x\}$. Then

$$\widehat{\rho}(x) := j(x,\rho(x))$$

is a rate of convergence for (γ_n) towards 0.
Conversely, for any rate ρ of convergence for (γ_n) we have that $\rho_e(x) := \rho(e+x)$ is a rate of convergence of $(\gamma(e)_n)_n$. In particular, the 2^{-e-x}-Cauchy property for (γ_n) implies the 2^{-x}-Cauchy property of $(\gamma(e)_n)_n$. Hence by Corollary 2.39, (γ_n) does not have an effectively learnable Cauchy rate. □

2.4 When learnability implies fluctuation bounds

In some cases, the effective (B, L)-learnability of a convergence rate (meaning that the convergence rate can be learned by L with $B(\underline{a})$-many mind changes) gives a bound on the number of fluctuations. This, for instance, is the case for bounded monotone sequences. In general, we can say that effective learnability implies the existence of an effective bound of fluctuations, if the learner and the sequence satisfy certain gap conditions. Informally, if

- any two exceptions $i, \tilde{\imath}$ to the Cauchy property for 2^{-k} have distance at least $\Delta_*(\max(i, \tilde{\imath}), k)$, and
- the learning map jumps at most by $J^*(i, k)$ from i, and
- (for any k) $J^*(\cdot, k)$ is asymptotically at most equivalent to $\Delta_*(\cdot, k)$,

then the number of fluctuations is asymptotically bounded by B^*. Below, we discuss an example from ergodic theory, where these conditions are met.

Proposition 2.41 (Gap conditions on the learner). *Let $a_{(\cdot)}$ be some sequence in a metric space (X, d) and let*

$$\varphi :\equiv \forall k \exists n \forall i \overbrace{\forall \tilde{\imath} < i \big(n \leq \tilde{\imath} \to d(a_{\tilde{\imath}}, a_i) \leq 2^{-k}\big)}^{\varphi_0(i,n,k):\equiv}.$$

be a (B, L)-learnable formula (which states simply the Cauchy property of the sequence, we use $\tilde{\imath}$ simply because the natural choice j already denotes the pairing function) and let B^*, L^* be majorants of B, L.
Moreover, suppose that there are functions $\Delta_* > 0$, J^*, s.t.

$$\forall n, i, i' \Big((\neg\varphi_0(i, n, k) \wedge \neg\varphi_0(i', n, k)) \to |i' - i| \geq \Delta_*(i, k) \Big),$$

$$\forall n, i (\neg\varphi_0(i, n, k) \to L^*(i, k) - i \leq J^*(i, k)),$$

and

$$(J^* \in \mathcal{O}(\Delta_*)) \equiv \forall k \exists N_k, K_k \, \forall x \geq N_k \, (K_k \Delta_*(x, k) \geq J^*(x, k)). \tag{20}$$

Then there is a bound on the number of 2^{-k}-fluctuations, which is primitive recursive in B^*, Δ_*, J^* and N_k, K_k (which witness (20)) given by

$$b(k) := B^*(k) \left(2 + K_k + \max_{n < N_k} \left(\frac{J^*(n, k)}{\Delta_*(n, k)} \right) \right).$$

Note that in the case where $N_k = 0$, we get

$$b(k) = (2 + K_k) B^*(k).$$

Proof. For simplicity, assume $N_k = 0$ (otherwise we could just replace every occurrence of K_k by $(K_k + \max_{n<N_k}(\frac{J^*(n,k)}{\Delta_*(n,k)}))$).
Firstly, note that any 2^{-k}-fluctuation between two indexes $\tilde{\imath}$ and i corresponds to a counterexample i. So, by definition there is at most one fluctuation in the interval $[c_l, i_l]$, where i_l is the smallest counterexample to the solution candidate c_l. Moreover, if there

is such a fluctuation, its greater index is i_l.
Secondly, from our assumption on Δ_* we get that there are at most

$$\left\lceil \frac{c_{l+1} - i_l}{\Delta_*(i_l, k)} \right\rceil \leq \left\lceil \frac{J^*(i_l, k)}{\Delta_*(i_l, k)} \right\rceil \leq K_k$$

many fluctuations within an interval $[i_l, c_{l+1}]$.
There are at most $B^*(k)$ such pairs of intervals, before a 2^{-k}-Cauchy point is reached, but there might be fluctuations, which arise only when we unite two such intervals.
By incrementing K_k by 1, we already account for any additional fluctuation due to combining the intervals $[c_l, i_l]$ and $[i_l, c_{l+1}]$. This is because if there was a fluctuation within $[c_l, i_l]$, then there cannot be an additional one which results from combining such a pair of intervals, as its greater index would be i_l. There can, however, be an additional fluctuation, when we combine the intervals $[i_l, c_{l+1}]$ and $[c_{l+1}, i_{l+1}]$.

□

We now consider the general form of the structure of Birkhoff's proof of the mean ergodic theorem as analyzed in [80] and the argument used in [10] to convert the rate of metastability obtained in [80] into a bound on the number of fluctuations:
Let $x_{(\cdot)}$ be a sequence in some normed space X (in the case at hand X is a uniformly convex Banach space) and $y_{(\cdot)}$ be a sequence in $\mathbb{R}_+$ definable by terms in $\mathsf{HA}^\omega[X, \|\cdot\|, \ldots]$. Suppose the Cauchyness of $x_{(\cdot)}$ is proved using that $y_{(\cdot)}$ has arbitrarily good approximate infima, i.e.

$$\forall \delta > 0 \exists n \forall k \forall \tilde{k} \leq k \, (y_{\tilde{k}} \geq y_n - \delta) \tag{21}$$
$$\rightarrow \forall \epsilon > 0 \exists m \forall u \forall i, j \in [m, u] \, (\|x_i - x_j\| \leq \epsilon). \tag{22}$$

This implication is classically equivalent to

$$\forall \epsilon > 0 \exists \delta > 0 \Big(\exists n \forall k \forall \tilde{k} \leq k \, (y_{\tilde{k}} \geq y_n - \delta) \rightarrow$$
$$\exists m \forall u \forall i, j \in [m, u] \, (\|x_i - x_j\| \leq \epsilon) \Big). \tag{+}$$

Suppose now that we are in the situation of Corollary 2.15, i.e.

$$\mathsf{HA}^\omega[X, \|\cdot\|, \ldots] + \mathrm{AC}{+}\mathrm{M}^\omega{+}\mathrm{IP}^\omega_\forall \vdash (+)$$

then

$$\mathsf{HA}^\omega[X, \|\cdot\|, \ldots] + \mathrm{AC} + \mathrm{M}^\omega + \mathrm{IP}^\omega_\forall \ \vdash$$
$$\forall \epsilon > 0 \exists \delta > 0 \forall n \exists m \geq n \forall u \exists k (\forall \tilde{k} \leq k (y_{\tilde{k}} \geq_{\mathbb{R}} y_n - \delta) \rightarrow \forall i, j \in [m, u] (\|x_i - x_j\| <_{\mathbb{R}} \epsilon).$$

Hence by monotone functional interpretation one extracts terms $\delta_\epsilon > 0$, m_ϵ and k_ϵ (depending additionally only on majorants of the parameters $\underline{a}$ used in the definition of our sequences) s.t. (valid in $\mathcal{S}^{\omega, X}$) for all majorants $\underline{a}^*$ of $\underline{a}$

$$\forall \epsilon > 0 \, \forall n, u \Big(m_\epsilon(n) \geq n \, \wedge$$
$$\left(\forall \tilde{k} \leq k_\epsilon(n, u) \, (y_{\tilde{k}} \geq y_n - \delta_\epsilon) \rightarrow \forall i, j \in [m_\epsilon(n), u] \, (\|x_i - x_j\| \leq \epsilon) \right) \Big). \tag{*}$$

Now define $k^*_\epsilon(u) := \max\{k_\epsilon(i,u) \ : \ i \le u\}$ and consider

$$\forall \epsilon > 0 \, \forall n, u \left(\forall \tilde{k} \le k^*_\epsilon(u) \, (y_{\tilde{k}} \ge y_n - \delta_\epsilon) \to \forall i, j \in [m_\epsilon(n), u] \, (\|x_i - x_j\| \le \epsilon) \right). \quad (**)$$

We can infer $(**)$ from $(*)$ by the following case distinction:[16] Fix $\varepsilon > 0$ and n.

Case 1: $u < m_\epsilon(n)$. Then the conclusion and hence the whole implication is trivially true.

Case 2: $u \ge m_\epsilon(n) \ge n$.Then $k^*_\epsilon(u) \ge k_\epsilon(n,u)$ and so $\forall \tilde{k} \le k^*_\epsilon(u) \, (y_{\tilde{k}} \ge y_n - \delta_\epsilon)$ implies $\forall \tilde{k} \le k_\epsilon(n,u) \, (y_{\tilde{k}} \ge y_n - \delta_\epsilon)$ and so the claim follows as well.

Now suppose w.l.o.g. that $k^*_\epsilon : \mathbb{N} \to \mathbb{N}$ is injective and for any given u define

$$l_u \ := \ (k^*_\epsilon)^{-1}(u).$$

Then $(**)$ applied to $u := l_u$ yields

$$\forall \epsilon > 0 \, \forall n, u \left(\forall \tilde{k} \le u \, (y_{\tilde{k}} \ge y_n - \delta_\epsilon) \to \forall i, j \in [m_\epsilon(n), (k^*_\epsilon)^{-1}(u)] \, (\|x_i - x_j\| \le \epsilon) \right). \ (\text{-})$$

Now let N_0, N_1, ..., N_{S_ϵ} be integers s.t. $N_0 = 0$ and N_{i+1} is the least $m > N_i$ s.t. $y_m < y_{N_i} - \delta_\epsilon$ as long as such an m exists. Assume that $b \ge y_0$ (for some b) and so $S_\epsilon \le \frac{b}{\delta_\epsilon}$.

By (-) there are no ϵ-fluctuations of $x_{(\cdot)}$ on the S_ϵ many intervals $[m_\epsilon(N_i), (k^*_\epsilon)^{-1}(N_{i+1})]$ for $i = 0, \ldots, S_\varepsilon - 1$.

In the intervals $[(k^*_\epsilon)^{-1}(N_i), m_\epsilon(N_i)]$ for $i = 1, \ldots, S_\epsilon$ and $[0, m_\epsilon(N_0)]$ we have to show that if we have for any $N \in \mathbb{N}$ s many fluctuations indexed within $[(k^*_\epsilon)^{-1}(N), m_\epsilon(N)]$ (or in $[0, m_\epsilon(N_0)]$) each indexed by a pair of indexes (i,j) then the highest index of such fluctuation (j_s) has to be greater than (or equal to) some $\varphi_\epsilon(s,N)$, where φ_ϵ is such that

$$\exists \tilde{s} \forall n \, (\varphi_\epsilon(\tilde{s}, n) > m_\epsilon(n)).$$

Then, given such an $\tilde{s}$, we have at most

$$\frac{b}{\delta_\epsilon} + \tilde{s} \left(\frac{b}{\delta_\epsilon} + 1 \right)$$

many fluctuations.

In the case of Birkhoff's proof, the analysis in [80] and the discussion in [10] gives the following data used in [10]:

$$\delta_\epsilon := \frac{\epsilon^2}{512b}, \qquad m_\epsilon(n) := \left\lceil \frac{16b}{\epsilon} \right\rceil n,$$

$$(k^*_\epsilon)^{-1}(n) := \left\lfloor \frac{n}{2} \right\rfloor, \qquad \varphi_\epsilon(s,n) := (1 + \frac{\epsilon}{2b})^s n,$$

and so (for $\varepsilon < 2b$)

$$\tilde{s} \le \frac{4b \log \lceil \frac{16b}{\epsilon} \rceil}{\epsilon}.$$

[16] We are grateful to P. Oliva for pointing this out to us.

The function φ_ε results (see [10] for the calculation) from the fact that

$$\|x_{n+k} - x_k\| \leq 2n\|x\|/(n+k)$$

which is established already in Birkhoff's proof and which – for $n = 1$ – shows that (x_k) has a linear rate of asymptotic regularity.

Remark 2.42. Naturally, we could use the data, which led to the bound of $\tilde{s}$ above, also simply with Proposition 2.41 to obtain a similar fluctuation bound (which has the same structure in ϵ).

For the case of Halpern iterations (with scalar $1/(n+1)$) mentioned in the introduction, the analysis given in [77] yields (roughly) the following data for Hilbert spaces X (see [77] for the detailed definition of Θ_n):

$$\delta_\epsilon := \frac{\epsilon^4}{576(b+1)^4}, \qquad m_\epsilon(n) \approx \Theta_n(\frac{\epsilon^2}{4}) \approx n^2,$$
$$(k_\epsilon^*)^{-1}(n) := \left\lfloor \frac{n \cdot \epsilon}{3b^2} \right\rfloor.$$

Similar data are also obtained in the recent [84] which is based on the analysis of a different proof for the strong convergence of the Halpern iteration from [131].
However, now the rate of asymptotic regularity roughly is of order (see corollary 6.3 in [77])

$$\|x_{k+1} - x_k\| \leq \frac{b}{\sqrt{k}},$$

which does not lead to a linear (in n) $\varphi_\varepsilon(s,n)$ and even if it would, this would not suffice to dominate $m_\varepsilon(n)$. So as it stands, the analysis does not seem to yield any fluctuation bound for the Halpern iteration (x_k).

Proposition 2.43. *Given a bound B_ε on the number of fluctuations, there is an in B_ε (and the given data $(k_\varepsilon^*)^{-1}$ and m_ε and the majorants $\underline{a}^*$ of their parameters including ε)) primitive recursive φ_ϵ satisfying the conditions in the proof:*

$$\forall \varepsilon > 0 \forall s, N, i, j (i, j \in [(k_\epsilon^*)^{-1}(N), m_\epsilon(N)] \wedge \mathrm{Fluc}_\varepsilon(s,i,j) \rightarrow j_s \geq \varphi_\epsilon(s,N)), \tag{23}$$
$$\exists \tilde{s} \forall n \, (\varphi_\epsilon(\tilde{s}, n) > m_\epsilon(n)). \tag{24}$$

Proof. Set

$$\varphi_\epsilon(s,n) := \begin{cases} (k_\epsilon^*)^{-1}(n) & \text{if } s \leq B_\varepsilon, \\ m_\epsilon(n) + 1 & \text{else.} \end{cases}$$

□

Remark 2.44. In particular, this means that if we know there is for computable (in the majorants $\underline{a}^*$ of the parameters including ε) $(k_\varepsilon^*)^{-1}$ and m_ε no computable φ_ε (in $\underline{a}^*$) satisfying these conditions, then there cannot be a bound on the number of fluctuations, which is computable (in $\underline{a}^*$).

Interpreting a strong non-linear ergodic theorem

This paper represents a significant contribution to the program of "proof mining" in nonlinear analysis, which aims to extract effective information hidden in the proofs. Applying methods coming from mathematical logic to results obtained by Wittmann [129], the author extracts effective rates of metastability (in the sense of Terence Tao) for nonlinear generalizations of the classical von Neumann mean ergodic theorem.
... the author gets a finitary version of the mean ergodic theorem for nonlinear mappings $T : X \to X$ satisfying $\|Tx + Ty\| \leq \|x + y\|$ for all $x, y \in X$.

– anonymous referee about [107], Dec 9, 2011.

Context of Proof Mining in Ergodic Theory

The Riesz version of the von Neumann mean ergodic theorem [128] asserts that for any linear operator T on a Hilbert space X, which is nonexpansive, i.e.

$$\forall u, v \in X \left(\|Tu - Tv\| \leq \|u - v\|\right),$$

the sequence of the Cesàro means

$$A_n x := \frac{1}{n+1} \sum_{i=0}^{n} T^i x,$$

converges in norm for any starting point x. It follows from an example by Genel and Lindenstrauss [31] that there is a nonexpansive operator on the unit ball of ℓ_2, for which the sequence of the Cesàro means does not converge strongly (see also [87]). So in comparison with von Neumann's linear mean ergodic theorem, in nonlinear ergodic theory one obtains either a weaker conclusion (such as weak convergence or convergence of a different iteration scheme instead of the Cesàro means) or one has to add additional requirements (to preserve at least some linearity).
Let H be a Hilbert space, C a subset of H and $T : C \to C$ a (possibly nonlinear) mapping. In 1975, Baillon [11] showed that if C is convex and closed, and T is nonexpansive

and has a fixed point, then the sequence of the Cesàro means is weakly convergent to a fixed point of T. A year later, Baillon [12] also proved that if in addition T is *odd*, i.e.

$$-C = C \text{ and } \forall u \in C\ \big(T(-u) = -Tu\big),$$

then the sequence of the Cesàro means converges to a fixed point in norm. Shortly after this, Brézis and Browder [18] showed that Baillon's first result is also true for a more general averaging process than the usual Cesàro means and that Baillon's second result remains valid if $0 \in C$ and T is not necessarily odd but satisfies the following, weaker condition:

$$\exists c \in \mathbb{R}\ \forall u,v \in C\big(\|Tu+Tv\|^2 \le \|u+v\|^2 + c(\|u\|^2 - \|Tu\|^2 + \|v\|^2 - \|Tv\|^2)\big). \quad \text{(BB)}$$

On the other hand, in 1979, Hirano and Takahashi [49] showed that Baillon's weak convergence result remains true if the mapping is just *asymptotically* nonexpansive, i.e.

$$\forall u,v \in C\ \forall n \in \mathbb{N}\ \big(\|T^n u - T^n v\| \le \alpha_n \|u - v\|\big),$$

for some sequence $(\alpha_n)_n$ of nonnegative real numbers which converges to 1. Moreover, an odd and nonexpansive mapping satisfies the following condition

$$\forall u,v \in C\ \big(\|T^n u + T^n v\| \le \|u+v\|\big) \quad (W)$$

and analogously an odd and asymptotically nonexpansive mapping satisfies the asymptotic version

$$\forall n \in \mathbb{N}\ \forall u,v \in C\ \big(\|T^n u + T^n v\| \le \alpha_n \|u+v\|\big), \quad (W^-)$$

for some sequence $(\alpha_n)_n$ of nonnegative real numbers which converges to 1. In 1990, Wittmann [129] proved a generalization of Baillon's strong convergence theorem to an arbitrary C and a mapping satisfying the condition (W^-) (see also Theorem 2.2 in [129] and Theorem 3.2 below). Two years later, Wittmann [130] also showed that for a nonexpansive T which has a fixed point, and a convex and closed C, the averaging sequence (x_n), first defined by Halpern [41] (for $x = 0$) as

$$x_0 := x, \quad x_{n+1} := \alpha_{n+1} x + (1 - \alpha_{n+1}) T(x_n),$$

converges in norm to the closest fixed point of T. The Halpern iteration coincides with the Cesàro means for linear maps and $\alpha_n = \frac{1}{n+1}$. We depict this development in Figure 1 (the references in parentheses refer to quantitative versions of the respective theorems, which we discuss below).

There are many further results and generalizations in the field of nonlinear ergodic theory (regarding different spaces see e.g. [21, 48], even weaker "linearity" conditions see e.g. [94, 105], and other improvements) and it is subject to ongoing research.

In this chapter we investigate the computational content of Wittmann's nonlinear strong ergodic theorem:

Theorem 3.1 (Wittmann 1990, [129]). *Let S be a subset of a Hilbert space and $T : S \to S$ be a mapping satisfying*

$$\forall x,y \in S\ \big(\|T^n x + T^n y\| \le \alpha_n \|x+y\|\big),\ \lim_{n\to\infty} \alpha_n = 1.$$

[19] While the results were essentially available on arxive since 2007, the paper as such was submitted in 2008. Thereafter Kohlenbach and Leuştean extended the result to uniformly convex Banach spaces and gave a better bound.

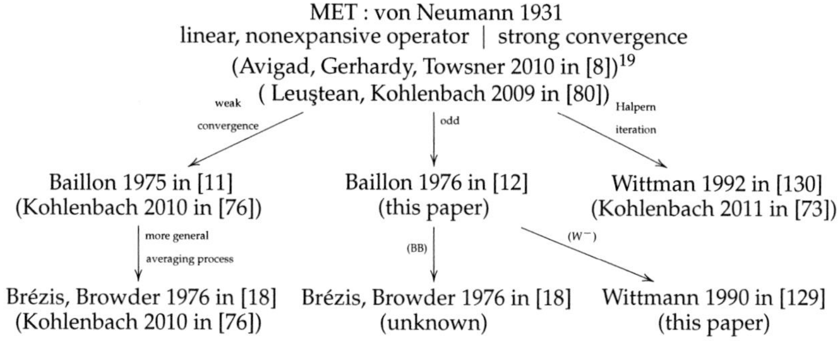

Figure 1: Some nonlinear ergodic theorems (for Hilbert spaces) and their finitisations.

Then for any $x \in S$ the sequence of the Cesàro means,

$$A_n x := \frac{1}{n+1} \sum_{i=0}^{n} T^i x,$$

is norm convergent.

Although in general the sequence of the ergodic averages does not have a computable rate of convergence (even for the von Neumann's mean ergodic theorem for a separable space and computable x and T), as was shown by Avigad, Gerhardy and Towsner in [8], the metastable version nevertheless has a primitive recursive bound. In this case it means that given the assumptions from Wittmann's strong ergodic theorem, the following holds

$$\forall b, l \in \mathbb{N}, g : \mathbb{N} \to \mathbb{N}, x \in S \; \exists m \leq M(l, g, b, K)\big(\|x\| \leq b \to \|A_m x - A_{m+g(m)} x\| \leq 2^{-l}\big),$$

for a primitive recursive M, where K is a rate of convergence for the sequence α in the assumption (W^-). We will not only prove the existence of such an M but also give such a bound explicitly in Corollary 3.12. For the specific case where $(\alpha_n)_n$ is a constant 1 sequence (i.e. T satisfies (W) rather than (W^-), see also Theorem 2.2 in [129] and Corollary 3.14 below), M can be defined as follows:

$$\begin{aligned}
M(l, g, b) &:= (N(2l+7, g^M, b) + P(2l+7, g^M, b))b2^{2l+8} + 1, \\
P(l, g, b) &:= P_0(l, F(l, g, N(l, g, b), b), b), \\
F(l, g, n, b)(p) &:= p + n + \tilde{g}((n+p)b2^{l+1}), \\
N(l, g, b) &:= \big(H(l, g, b)\big)^{b^2 2^{l+2}}(0), \\
H(l, g, b)(n) &:= n + P_0(l, F(l, g, n, b), b) + \tilde{g}((n + P_0(l, F(l, g, n, b), b))b2^{l+1}),
\end{aligned}$$

where (for $f : \mathbb{N} \to \mathbb{N}$)

$$P_0(l, f, b) := \tilde{f}^{b^2 2^l}(0), \quad \tilde{g}(n) := n + g(n), \; g^M(n) := \max_{i \leq n+1} g(i).^{20}$$

Note that apart from the counterfunction g and the precision l, this bound depends only on b and not on S, T or x. For another quantitative result on operators satisfying the condition (W) see [74].

These results, along with those by Avigad, Gerhardy, Towsner [8] and Kohlenbach, Leuştean [80] for the finitary version of the von Neumann ergodic theorem as well as Kohlenbach's bounds for the finitary versions of Baillon's weak ergodic theorem [76] and Wittmann's convergence result for Halpern means [73] , can be seen as instances of 'hard analysis' in the sense of T. Tao; see [123, 122], where he discusses the uses and benefits of (the existence of) uniform bounds for such finitary formulations of well-known theorems. It is one of the goals of this chapter to demonstrate that there are proof-theoretic means to systematically obtain such uniform bounds. In fact, for many theorems the existence of a uniform bound is guaranteed by Kohlenbach's metatheorems introduced in [70] and refined in [37]. Additionally, proof theoretic methods such as Kohlenbach's monotone functional interpretation (see [62]) can be used to systematically obtain these effective bounds. This chapter is a case study in applying such proof mining techniques.

We improve results in the area of nonlinear generalizations of the mean ergodic theorem and their corresponding finitisations (see Figure 1). Moreover, we have here a rare example of an application of these techniques to not necessarily continuous operators. In logical terms this amounts to the subtlety that only a weak version of extensionality is available. Also, for the first time, we obtain a bound which in fact makes use of nested iteration. One can see this quickly on the term M above. While F as a function is defined via iteration of the counterfunction g, it itself is being iterated by P. This is a direct consequence of the logical form of Wittmann's original proof.

It is a surprising observation that so far for all metastable versions of strong ergodic theorems primitive recursive bounds could be obtained.

We discuss the application of general logical metatheorems in more detail in Section 3.3 which is not necessary to understand and verify our main results. We present these, namely the explicit bounds for all three theorems in Wittmann's paper [129], in Section 3.4. We explain how to obtain the explicit bounds in Section 3.2, after formalizing Wittmann's proof in Section 3.1. Both of these sections, though inspired by proof theoretic methods, require no facts from logic.

3.1 Arithmetizing Wittmann's proof

The first step of a proof mining process is to investigate the proof of the theorem we want to analyze. In addition to chapter 1, in particular for a discussion on proof mining techniques in connection with ergodic theory see [35] and the last section of [8]. For an in-depth analysis of applied proof theory see [71]. The nonconstructive, or ineffective, content of Wittmann's proof are the principle of convergence for bounded monotone sequences of real numbers and the existence of infimum for bounded sequences of real numbers. Formulated in the usual way, both principles state the existence of a real number, which we represent as fast converging Cauchy sequences of rationals encoded as number theoretic functions (i.e. functions in $\mathbb{N}^{\mathbb{N}}$, see Section 1.4). However, for a given sequence, both principles can be replaced by weaker statements about natural numbers only (as opposed to statements about objects in $\mathbb{N}^{\mathbb{N}}$). In the presence of arithmetical comprehension, these weaker (arithmetical) statements are equivalent to the original (analytical) principles.[21] For the convergence we work with the arithmetic

[21] While the analytical principles are actually known to be equivalent to arithmetic comprehension (see [115] and – for more detailed results – [68]), the arithmetic versions are equivalent to Σ^0_1-induction and

Cauchy property and for infimum we give for any precision an approximate infimum.

1. Arithmetized convergence of a monotone bounded sequence $a_{(\cdot)}$:

$$\forall l \exists n \forall m \geq n \quad \big(|a_n - a_m| \leq 2^{-l}\big).$$

2. Arithmetized existence of the infimum of a bounded sequence $a_{(\cdot)}$:

$$\forall l \exists n \forall m \quad \big(a_n - a_m \leq 2^{-l}\big).$$

Of course, in this way we don't get a single point which *is* the limit point or infimum. Therefore we have to analyze the proof and see whether such points are actually needed or whether these arithmetical versions suffice. Here, fortunately, it turns out that the latter is the case (see [65] for a general discussion of this point).
Following [71], we show that we can rewrite Wittmann's proof (see [129]) in the language of $\mathcal{A}^\omega[X, \langle\cdot,\cdot\rangle]$, carefully using only weak (arithmetized) principles at the relevant places.

Theorem 3.2 (Wittmann 1990, [129]). *Let $X_{(\cdot)}$ be a sequence in a Hilbert space s.t. for all $m, n, k \in \mathbb{N}$*

$$||X_{n+k} + X_{m+k}||^2 \leq ||X_n + X_m||^2 + \delta_k \tag{i}$$

with

$$\lim_{k\to\infty} \delta_k = 0. \tag{ii}$$

Then the sequence $A_{(\cdot)}$ defined by

$$A_n := \frac{1}{n}\sum_{i=1}^{n} X_i,$$

is norm convergent.

We follow Wittmann's notation and use $X_{(\cdot)}$ to denote the sequence in the Hilbert space (not to be confused with the Hilbert space itself, which might be implied by the notation $\mathcal{A}^\omega[X, \langle\cdot,\cdot\rangle]$). Also, to keep the proof more readable, we refer the more technical steps to later sections. There we need to do a thorough analysis of those steps to obtain the precise bounds of the realizers, while here we can settle for their existence.
We formulate conditions (i) and (ii) from the theorem as arithmetical statements as follows (except for the variable K, which we will treat as a given parameter):

$$\forall m, n, k \in \mathbb{N} \quad \big(||X_{n+k} + X_{m+k}||^2 \leq ||X_n + X_m||^2 + \delta_k\big), \tag{25}$$

$$\exists K \in \mathbb{N}^{\mathbb{N}} \forall n \in \mathbb{N} \forall i \geq K(n) \quad \big(|\delta_i| \leq 2^{-n}\big). \tag{26}$$

From now on, let K always denote a rate of convergence of the sequence $\delta_{(\cdot)}$ (i.e. a function satisfying (A2)) and B the upper bound of $X_{(\cdot)}$ (such a bound can be defined primitive recursively in $K(0)$ and some elements of $X_{(\cdot)}$, see Proposition 3.9).
It is easy to show that the sequence $(||X_n||)_n$ is a Cauchy sequence (see proof of Lemma 3.19 below). Such an arithmetical formulation of the convergence of $(||X_n||)_n$ is sufficient to infer that in particular we have that

$$\forall l^0 \exists n^0 \forall m_2^0 \geq m_1^0 \geq n \quad \big(\, ||X_{m_1}||^2 \leq ||X_{m_2}||^2 + 2^{-(l+1)}\,\big). \tag{27}$$

hence have a functional interpretation by ordinarily primitive recursive functionals (see [71]).

The Skolem function realizing n^0 would correspond to n_ϵ in Wittmann's proof (where he used the standard convergence), however we work only with the fact that such an n exists for any given precision l. From 27 we infer that

$$\forall l \exists n_l \forall m, n \geq n_l \, \forall k \geq K(l) \quad (\langle X_{n+k}, X_{m+k} \rangle \leq \langle X_n, X_m \rangle + 2^{-l}), \tag{W1}$$

the same way as Wittmann in [129], see also the end of the proof of Lemma 3.21. From now on, by n_l we always denote an n_l which satisfies (W1).

Since the norm of any convex combination of the elements of the sequence is bounded from below by 0, the set of all such convex combinations has an infimum. To give an arithmetic formulation of this fact, it is useful to have a primitive recursive functional C which gives us a 2^{-l} approximation of the smallest convex combination of $X_n, X_{n+1}, \ldots, X_p$ (more precisely, of the square of the infimum of the set of the norms of these convex combinations). Clearly, there are many ways to define such a functional. We do so in a straightforward way in Definition 3.16 below. Having such a C in place, the following arithmetical formula

$$\forall l \exists p_l \forall p \geq p_l \quad (C(l, n_l, p_l) \leq C(l, n_l, p) + 2^{-l}), \tag{W2}$$

states the existence of an approximate infimum of all convex combinations of $X_{(\cdot)}$ (see also Lemma 3.17). Similarly as with n_l, from now on by p_l we mean a number satisfying (W2).

Wittmann introduces a specific point $z_{\epsilon,m}$, which we denote by $Z(l, n, p, m)$ and define using our notation in Definition 3.22 below. However, here the precise definition is not as important as the properties of this point. Firstly, we show that for any two natural numbers i and j, the distance between $Z(l, n_l, p_l, i)$ and $Z(l, n_l, p_l, j)$ is arbitrary small for sufficiently large l. Together with the convexity of the square function we have (again, see [129] or the proof of Lemma 3.23.(2) below) that

$$\forall l^0, m^0 \quad (\|Z(l, n_l, p_l, m)\|^2 \leq C(l, n_l, p_l) + 2^{-l}). \tag{W3}$$

From (W2) we can infer (see the proof of Lemma 3.23.(1) below) that

$$\forall l, i, j \left(\left\|\frac{1}{2}(Z(l, n_l, p_l, i) + Z(l, n_l, p_l, j))\right\|^2 + 2^{-l} \geq C(l, n_l, p_l) \right),$$

and together with (W3) and the parallelogram identity we can conclude that

$$\forall l, i, j \left(\|Z(l, n_l, p_l, i) - Z(l, n_l, p_l, j)\|^2 \leq 2^{-l+4} \right).$$

Secondly, it follows from the definition of Z (yet again see [129] or proof of Lemma 3.24) that

$$\forall l \forall m \geq p_l \left(\|Z(l, n_l, p_l, m) - A_{m+1}\| \leq \frac{2(p_l + n_l + K(l)) \sup_{n \in \mathbb{N}} \|X_n\|}{m+1} \right),$$

which means that we can make the distance between A_{m+1} and $Z(l, n_l, p_l, m)$ arbitrarily small by choosing m sufficiently large.

In particular we have shown that the distance between any A_i and any A_j is arbitrarily small once i and j are sufficiently large. Note that we can choose an arbitrarily large l first and then we are still free to choose sufficiently large i and j after n_l and p_l are fixed. This concludes the sketch of the proof that $A_{(\cdot)}$ is a Cauchy sequence.

3.2 Obtaining a bound

The goal of this section is to roughly describe how to find a bound for m in

$$\forall l, g \exists m \left(\|A_m - A_{m+g(m)}\| \leq 2^{-l} \right).$$

Similarly as before, for better understanding we leave some technical details for later sections and disregard the monotonicity of the bounds as well as some small corrections needed in the exponent of 2^{-l}. We handle these two aspects very carefully in the sections 3.4 and 3.5, where we make also all of the following steps more explicit.
Let the assumptions in the proof of Theorem 3.2 hold and let K, B and C be as before as well.
Furthermore, we assume that N_0 and P_0 are the witnessing terms for (W1) (note that in fact this means rather that we have a witness for Kreisel's no-counterexample interpretation – n.c.i. see [85, 86] and Remark 1.21 – of the convergence of $\|X_n\|$ in the first place) and (W2) as given in [71], i.e. we have that:

$$\forall l, h \forall m, n \in [N_0(l,h); N_0(l,h) + g(N_0(l,h))] \left(\langle X_{m+k}, X_{n+k} \rangle \leq \langle X_m, X_n \rangle + 2^{-l} \right)$$

and

$$\forall l, f \left(C(l, n_l, P_0(l,f)) \leq C(l, n_l, f(P_0(l,f))) + 2^{-l} \right)$$

for any n_l satisfying (W1).
This structure accounts for the already mentioned nested iteration. Eventually, we have to specify a counterfunction f s.t. it is sufficient that this inequality holds for that particular f. This f has to depend on n_l, which will obviously be defined as an iteration, and f itself will be iterated by P_0.
To obtain a bound for m, we follow the proof from the last section backwards. We define

$$M_0(l, n, p) := 2(p + n + K(l))B2^l,$$

since then we have that $\|Z(l, n_l, p_l, M_0(l, n_l, p_l)) - A_{m+1}\| \leq 2^{-l}$, for the right values of l, n_l and p_l (which we don't know yet).
From the proof we can infer that the largest p needed in (W2) is $K(l) + m + g(m) + n_l + p_l$ (for details see proof of Lemma 3.23.(1)), therefore we need that

$$f(P_0(l,f)) = K(l) + m + g(m) + n_l + P_0(l,f),$$

(where m and $m + g(m)$ correspond to the m and n in (W1)).
Moreover, we can see that the largest m or n needed in (W1) is $n_l + p_l + m + g(m) + K(l)$ (for details see proof of Lemma 3.21 and its application in proof of Lemma 3.23.(2)). So we need that $h(N_0(h,l)) = p_l + m + g(m) + K(l)$.
Keeping in mind that $N_0(l,h)$ corresponds to n_l (and to n below) and $P_0(l,f)$ to p_l (and to p below) we obtain[22]

$$h(l,g) = \lambda n \, . \, \underbrace{P_0(l, \lambda p \, . \, K(l) + M_0(l,n,p) + g(M_0(l,n,p)) + p + n)}_{p_l} + \\ + M_0(p_l, n, l) + g(M_0(p_l, n, l)) + K(l),$$

and define

$$N(l,g) := N_0(l, h(l,g)).$$

[22] Here and in the following, when considering a complex term $t[a_1, \ldots, a_n]$ in several variables $a_1, \ldots, a_n$, the notation $\lambda a_i . t[a_1, \ldots, a_n]$ defines (as usual in logic) the function $a_i \mapsto t[a_1, \ldots, a_n]$.

Given $N(l,g)$, which corresponds to n_l, and again keeping in mind that $P_0(l,f)$ corresponds to p_l (and to p below), we can define $P(l,g)$ as well:

$$P(l,g) := P_0(l, f(l,g)),$$

where $f(l,g) = \lambda p \,.\, K(l) + M_0(p, N(l,g), l) + g(M_0(p, N(l,g), l)) + p + N(l,g)$. Finally we can define the desired witness for m as follows:

$$M(l,g) := M_0(l, N(l,g), P(l,g)).$$

3.3 A general bound existence theorem

The main result of this paper, Corollary 3.12, is a quantitative version of a nonlinear strong ergodic theorem for operators satisfying Wittmann's condition (W^-) on an arbitrary subset of a Hilbert space. In this section we outline how for this type of theorems the existence of such uniform bounds can be obtained by means of a general logical metatheorem. The metatheorem applicable in our scenario follows from Corollary 6.6.7) in [37] and we introduced it in the first chapter (Theorem 1.33). However, let us recall it here, for reader's convenience.

Theorem 3.3 (Gerhardy-Kohlenbach [37] - specific case 1). *Let $\varphi_\forall$, resp. $\psi_\exists$, be $\forall$- resp. $\exists$-formulas that contain only x, z, f free, resp. x, z, f, v free. Assume that $\mathcal{A}^\omega[X, \langle\cdot,\cdot\rangle, S]$ proves the following sentence:*

$$\forall x \in \mathbb{N}^{\mathbb{N}}, z \in S, f \in S^S \left(\varphi_\forall(x,z,f) \to \exists v \in \mathbb{N}\, \psi_\exists(x,z,f,v)\right).$$

Then there is a computable functional $F : \mathbb{N}^{\mathbb{N}} \times \mathbb{N} \times \mathbb{N}^{\mathbb{N}} \to \mathbb{N}$ s. t. the following holds in all non-trivial (real) inner product spaces $(X, \langle\cdot,\cdot\rangle)$ and for any subset $S \subseteq X$

$$\forall x \in \mathbb{N}^{\mathbb{N}}, z \in S, b \in \mathbb{N}, f \in S^S, f^* \in \mathbb{N}^{\mathbb{N}}$$
$$\left(\mathrm{Maj}(f^*, f) \wedge \|z\| \le b \wedge \varphi_\forall(x,z,f) \to \exists v \le F(x,b,f^*)\, \psi_\exists(x,z,f,v)\right),$$

where

$$\mathrm{Maj}(f^*, f) :\equiv \forall n \in \mathbb{N} \forall z \in S \left(\|z\| \le_{\mathbb{R}} n \to \|f(z)\| \le_{\mathbb{R}} f^*(n)\right).$$

The theorem holds analogously for finite tuples.

Consider the following metastable version of Wittmann's Theorem 2.1 [129]:

Theorem 3.4 (Theorem 2.1 in [129]). *Let S be a subset of a Hilbert space and $T : S \to S$ be a mapping satisfying*

$$\forall x, y \in S\ (\|Tx + Ty\| \le \|x + y\|). \tag{W}$$

Then for any $x \in S$ the sequence of the Cesàro means,

$$A_n x := \frac{1}{n+1} \sum_{i=0}^{n} T^i x,$$

is norm convergent.

This theorem has the following form:

$$\forall l \in \mathbb{N}, g \in \mathbb{N}^{\mathbb{N}}, x \in S, T \in S^S \left(W(T) \to \exists m \in \mathbb{N} \left(\|A_m x - A_{m+g(m)} x\| < 2^{-l} \right) \right). \quad (+)$$

Obviously the conclusion, i.e. $\exists m \left(\|A_m x - A_{m+g(m)} x\| < 2^{-l} \right)$, has the form $\exists m\, \psi_\exists(m, l, g)$ and the assumption $W(T)$, i.e. $\forall x, y \in S(\|Tx + Ty\| \le \|x + y\|)$, has the form $\varphi_\forall(T)$. Moreover, $W(T)$ already implies $\mathrm{Maj}(\mathrm{id}, T)$ (here id stands simply for the identity function on $\mathbb{N}$), since $W(T)$ applied to $x = y = z$ implies $\forall z \in S(\|T(z)\| \le \|z\|)$.
Hence we can apply Theorem 1.33 to (+) by setting

$$\underline{x} :=_{\mathbb{N} \times \mathbb{N}^{\mathbb{N}}} l, g, \; z :=_S x, \; f :=_{S \to S} T, \; f^* :=_{\mathbb{N} \to \mathbb{N}} \mathrm{id},$$

and

$$\varphi_\forall(x, z, f) :\equiv W(T), \; \exists v \in \mathbb{N}\, \psi_\exists(x, z, f, v) :\equiv \exists m \in \mathbb{N} \left(\|A_m x - A_{m+g(m)} x\| < 2^{-l} \right),$$

to obtain that there is a computable bound $M : \mathbb{N} \times \mathbb{N}^{\mathbb{N}} \times \mathbb{N} \to \mathbb{N}$, s.t.

$$\forall l \in \mathbb{N}, g \in \mathbb{N}^{\mathbb{N}}, x \in S, T \in S^S$$
$$\left(W(T) \wedge \|x\| \le b \;\to \exists m \le_{\mathbb{N}} M(l, g, b) \left(\|A_m x - A_{m+g(m)} x\| \le 2^{-l} \right) \right).$$

It is rather easy to see that the proof can be formalized in $\mathcal{A}^\omega[X, \langle \cdot, \cdot \rangle, S]$, except for the question of the use of the axiom of extensionality (full extensionality is in general unavailable in any proof-theoretic extraction of computational bounds). Generally, one can avoid the use of full extensionality in proofs of statements about continuous objects. Note that in particular any nonexpansive operator is also continuous. However, in our case, the operator T may be discontinuous. Fortunately, Wittmann proves his main results as a consequence of a statement about a simple sequence of elements in S, which as such is independent of T (see Theorem 2.3 in [129] or Theorem 3.6 below), whereby all relevant equalities are provable directly. Therefore the rule of extensionality suffices to formalize his proof.
Hence the existence of a *uniform computable bound* for the metastable version can be inferred from the metatheorem in [37]. Furthermore, since the metatheorem is established by proof-theoretic reasoning, it provides not only the existence of a uniform bound but also a procedure for its extraction.
As discussed in Section 1.15, in general such a bound might need bar-recursion (see Section 1.11), which is required to interpret the schema of full comprehension over numbers in Spector's system (see also [117]). However, once more due to the way how Wittmann proved the analyzed theorem, it is easy to see that the only proof-theoretically non-trivial principles needed in the proof are the existence of the infimum/supremum of bounded sequences and the principle of convergence for bounded monotone sequences. Both of these principles need only bar-recursion restricted to numbers and functions ($\mathsf{BR}_{0,1}$) and not full BR. (Kohlenbach shows in [71, 68] that both principles are provable from arithmetical comprehension which is interpreted in $\mathcal{T}_0 + \mathsf{BR}_{0,1}$.) Moreover, since the bound itself has only functions and numbers as arguments, it follows from [111, 67] that the bound is not only computable, but that the *bound is a primitive recursive functional in the sense of Gödel's $\mathcal{T}$*.
These observations can be made a priori, without any in-depth analysis of the proof. In addition, one more conclusion can be drawn before one actually extracts the bound. In general, it is helpful, and sometimes even necessary, to simplify (in the sense of

proof-theoretic strength) the analyzed proof. We do so in Section 3.1. As one can see, in fact the proof uses only arithmetical versions of the non-trivial principles (which can be proved by Σ^0_1-IA) and therefore we know that the use of $\mathsf{BR}_{0,1}$ can be eliminated as well. In fact, for these arithmetic versions the bounds for the witnesses for the metastable formulations are already known and rather simple.

Proposition 3.5 (Kohlenbach [71]). *Let (a_n) be a nonincreasing sequence in $[0,C]$ for some constant $C \in \mathbb{N}$, then*

$$\forall k \in \mathbb{N}, g \in \mathbb{N}^{\mathbb{N}} \exists n \leq F(g,k,C) \forall i,j \in [n; n+g(n)] \left(|a_i - a_j| < 2^{-k}\right),$$

where $F(g,k,C) := \tilde{g}^{C\cdot 2^k}(0)$ with $\tilde{g}(n) := n + g(n)$.

Proof. See Propositions 2.27 and Remark 2.29 in [71].

□

Hence, we can infer that there is actually *an ordinary primitive recursive bound* (a bound in $\mathcal{T}_0$) which we give explicitly in Section 3.4, Corollary 3.14.
We should point out that the original Corollary in [37] can be used in a more general context than the particular example we just discussed. For instance, it can be applied to both, Theorem 2.2 and Theorem 2.3 in [129]. Take for example the metastable formulation of Theorem 2.3 in [129]:

Theorem 3.6. *Let $X_{(\cdot)}$ be a sequence in a Hilbert space s.t.*

$$\forall m,n,k \in \mathbb{N} \quad \left(\|X_{n+k} + X_{m+k}\|^2 \leq \|X_n + X_m\|^2 + \delta_k\right),$$

and $\exists K \in \mathbb{N}^{\mathbb{N}} \forall n \in \mathbb{N} \forall i \geq Kn \quad (\delta_i \leq 2^{-n})$. Then

$$\forall l \in \mathbb{N}, g : \mathbb{N} \to \mathbb{N}\ \exists m \quad \left(\|A_m x - A_{m+g(m)} x\| \leq 2^{-l}\right).$$

For simplicity, let us here assume that the sequence $\delta_{(\cdot)}$ is in the real unit interval. In this case we have the following additional parameters: $K : \mathbb{N} \to \mathbb{N}$, $\delta : \mathbb{N} \to [0,1]$ and a sequence $z_{(\cdot)} := X_{(\cdot)}$ (rather than a starting point $z := x$). We can apply Corollary 6.6.7) in [37] for the theory $\mathcal{A}^\omega[X, \langle\cdot,\cdot\rangle, S]$ again. Additionally we use point 6.6.3). On the other hand, this time we don't need the function parameter f:

Theorem 3.7 (Gerhardy-Kohlenbach [37] - specific case 2). *Let P be a $\mathcal{A}^\omega$-definable Polish space and let $\varphi_\forall$, resp. $\psi_\exists$, be a $\forall$- resp. an $\exists$-formula that contains only x, z, f free, resp. x, z, f, v free. Assume that $\mathcal{A}^\omega[X, \langle\cdot,\cdot\rangle, S]$ proves the following sentence:*

$$\forall x \in \mathbb{N}^{\mathbb{N}}, y \in P, z_{(\cdot)} \in \mathbb{N}^S \left(\varphi_\forall(x,z) \to \exists v \in \mathbb{N}\ \psi_\exists(x,z,v)\right),$$

Then there is a computable functional $F : \mathbb{N}^{\mathbb{N}} \times \mathbb{N}^{\mathbb{N}} \to \mathbb{N}$ s. t. the following holds in all non-trivial (real) inner product spaces $(X, \langle\cdot,\cdot\rangle)$ and for any subset $S \subseteq X$

$$\forall x \in \mathbb{N}^{\mathbb{N}}, z_{(\cdot)} \in \mathbb{N}^S, b_{(\cdot)} \in \mathbb{N}^{\mathbb{N}}$$
$$\left(\forall n \in \mathbb{N}\ (\|z_n\| \leq b_n) \wedge \varphi_\forall(x,z) \to \exists v \leq F(x, b_{(\cdot)}) \psi_\exists(x,z,v)\right).$$

The theorem holds analogously for finite tuples.

Given any rate of convergence for the sequence $\delta_{(\cdot)}$, the metastable version of the assumptions in Theorem 3.6 is purely universal:

$$\forall m,n,k,j \in \mathbb{N}\, \forall i \geq K(j)\ \left(\|X_{n+k} + X_{m+k}\|^2 \leq \|X_n + X_m\|^2 + \delta_k \wedge \delta_i \leq 2^{-n}\right). \quad (\mathrm{W}')$$

Moreover, the Polish space $[0,1]^{\mathbb{N}}$ is naturally definable in $\mathcal{A}^\omega$ so we obtain by Theorem 3.7 that

$$\begin{aligned}\forall l \in \mathbb{N}, g \in \mathbb{N}^{\mathbb{N}}, K \in \mathbb{N}^{\mathbb{N}},\ \delta_{(\cdot)} \in [0,1]^{\mathbb{N}}, X_{(\cdot)} \in S^{\mathbb{N}}, b_{(\cdot)} \in \mathbb{N}^{\mathbb{N}} \\ \big(\, \forall i \in \mathbb{N}\ (X_i \leq b_i) \wedge \mathrm{W}'(X,K,\delta) \to \\ \exists m \leq_{\mathbb{N}} M(l,g,b_{(\cdot)},K)\ (\|A_m x - A_{m+g(m)} x\| \leq 2^{-l})\,\big).\end{aligned}$$

Similarly as before, W$'$ implies $\forall i \in \mathbb{N}\ (X_i \leq b_i)$ for a suitable $b_{(\cdot)}$. We give such a bound explicitly in Section 3.4. To be precise, we give a bound $M(l,g,b_0,b_1,\dots,b_{K(0)},K)$ for an arbitrary sequence $\delta_{(\cdot)} \in \mathbb{R}^{\mathbb{N}}$ converging to zero with the rate K. In the simplified case for the unit interval, it is straightforward to see that the bound also simplifies to an even more uniform bound $M(l,g,b_0,K)$.
To repeat, these are very specific scenarios. We should emphasize that the Corollaries in [37], and the metatheorem(s) even more so, have a much wider range of application.

3.4 Uniform bounds for Wittmann's ergodic theorems

We give a bound for a finitary version of Wittmann's convergence result for a general series in a Hilbert space satisfying a suitable formulation of the condition (W^-) first (see Proposition 3.9 and Theorem 3.11) to derive the bounds for the finitary versions of the actual ergodic theorems later (see Corollary 3.12 and Corollary 3.14).
The proof of Proposition 3.9 is rather involved and we give it in a dedicated section. We denote the set of natural numbers $\{x \in \mathbb{N} : a \leq x \wedge x \leq b\}$ by $[a;b]$.

Remark 3.8. In the following proposition, we omit the term dependencies whenever the arguments are trivial. In particular we omit the dependency on the parameters K and B in the definition of M'. E.g. we write $F(p) := (F(l,g,n))(p) := p + n + K^{\mathsf{M}}(l) + M_0 + g(M_0)$ to define a functional F, which given the natural numbers l, n and B and the functions $g : \mathbb{N} \to \mathbb{N}$ and $K : \mathbb{N} \to \mathbb{N}$ returns a function $F(l,g,n,K,B) : \mathbb{N} \to \mathbb{N}$ which maps any natural number p to the natural number $p + n + K^{\mathsf{M}}(l) + M_0(l,n,p,K,B) + g(M_0(l,n,p,K,B))$.

Proposition 3.9 (Finitary version of Theorem 2.3 in [129]). *Let $K : \mathbb{N} \to \mathbb{N}$ be a function and $X_{(\cdot)}$ a sequence in a Hilbert space s.t. for all $m,n,k \in \mathbb{N}$*

$$\|X_{n+k} + X_{m+k}\|^2 \leq \|X_n + X_m\|^2 + \delta_k, \quad (\mathrm{A1})$$

with

$$\forall l \in \mathbb{N}\forall n \geq K(l)\ (|\delta_n| < 2^{-l}), \quad (\mathrm{A2})$$

in other words let K be a rate of convergence of (δ_k) towards 0. Furthermore let B be a natural number s.t. $B \geq \|X_i\| + 1$ for all $i \leq K(0)$.
Then the sequence $A_{(\cdot)}$, defined by $A_n := \frac{1}{n}\sum_{i=1}^{n} X_i$, is a Cauchy sequence and we have that

$$\forall l \in \mathbb{N}, g : \mathbb{N} \to \mathbb{N}\ \exists m \leq M'(l,g^{\mathsf{M}},K,B)\ \ (\|A_m - A_{m+g(m)}\| \leq 2^{-l}),$$

with $g^{\mathsf{M}}(n) := \max_{i\le n} g(n)$, $K^{\mathsf{M}}(n) := \max_{i\le n} K(n)$, *and* $M'(l,g,K,B)$ *defined as follows:*

$$\begin{aligned}
M' &:= M'(l,g) := M(2l+6,g')+1,\ g':\mathbb{N}\to\mathbb{N},\ g'(n):=g(n+1),\\
M &:= M(l,g) := M_0(P,N,l),\\
P &:= P(l,g) := P_0(l,F(l,g,N)),\\
F(p) &:= (F(l,g,n))(p) := p+n+K^{\mathsf{M}}(l)+M_0+g(M_0),\ F:\mathbb{N}\to\mathbb{N},\\
N &:= N(l,g) := N_0(l+1,H),\\
H(n) &:= (H(l,g))(n) := H_0(l,g,n,P_0(l,F)),\ H:\mathbb{N}\to\mathbb{N},
\end{aligned}$$

where

$$\begin{aligned}
H_0 &:= H_0(l,g,n,p) := p+M_0+g(M_0)+K^{\mathsf{M}}(l),\\
M_0 &:= M_0(l,n,p) := (2n+2p+2K^{\mathsf{M}}(l))B2^l,\\
P_0 &:= P_0(l,f) := \tilde{f}^{B^2 2^l}(0),\quad \tilde{f}(n):=n+f(n),\\
N_0 &:= N_0(l,h) := P_0(l+1,U)+K^{\mathsf{M}}(l+1),\\
U(n) &:= (U(l,h))(n) := (n+K^{\mathsf{M}}(l+1))+h^{\mathsf{M}}(n+K^{\mathsf{M}}(l+1)),\ U:\mathbb{N}\to\mathbb{N}.
\end{aligned}$$

Remark 3.10. From now on we will make use of the following observations regarding this proposition.

1. Due to the condition (A1) we have that $\forall i\in\mathbb{N}\ B\ge\|X_i\|$.
2. The condition (A2) holds for K^{M} as well. Therefore it is safe to assume that K is already monotone and hence that $K^{\mathsf{M}}=K$.

Sometimes it is useful to work with the following version of the previous theorem, though both these formulations are equivalent (even in very weak systems).

Theorem 3.11 (Finitary version of Theorem 2.3 in [129] for intervals). *Given the assumptions in Proposition 3.9 the sequence* $A_{(\cdot)}$ *defined by*

$$A_n := \frac{1}{n}\sum_{i=1}^{n} X_i,$$

is a Cauchy sequence and we have that

$$\forall l\in\mathbb{N}, g:\mathbb{N}\to\mathbb{N}\ \exists m\le M'(l+1,g^{\mathsf{M}},K,B)\ \forall i,j\in[m;m+g(m)]\ \ (\|A_i-A_j\|\le 2^{-l}),$$

with $M'(l,g,K,B)$ *defined as in Proposition 3.9.*

Proof. Given any l and g, apply Proposition 3.9 to the number $l+1$ and to the function

$$h(n) := \min\Big\{\, i\in[0;g(n)] \text{ s.t. } \forall j\in[0;g(n)]\ \ \Big(\big|\|A_{n+i}\|-\|A_n\|\big|\ge\big|\|A_{n+j}\|-\|A_n\|\big|\Big)\Big\}.$$

It follows that (here again, we omit obvious dependencies on trivial arguments)

$$\exists m\le M'(l+1,h^{\mathsf{M}})\ (\|A_m-A_{m+h(m)}\|\le 2^{-l-1}).$$

We fix such an m and conclude (by the triangle inequality) that

$$\forall i,j \in [m; m+g(m)] \ (\|A_i - A_j\| \leq 2^{-l}).$$

Moreover, since $\forall n \in \mathbb{N} \ (h^{\mathsf{M}}(n) \leq g^{\mathsf{M}}(n))$ we have that

$$m \leq M'(l+1, g^{\mathsf{M}})$$

due to Lemma 3.18. □

Now, we obtain the bound for the metastable version of Theorem 2.2 in Wittmann's paper [129] as a simple conclusion:

Corollary 3.12 (Finitary version of Theorem 2.2 in [129]). *Let S be a subset of a Hilbert space and $T : S \to S$ be a mapping satisfying*

$$\forall n \in \mathbb{N} \forall x, y \in S \ (\|T^n x + T^n y\| \leq \alpha_n \|x + y\|), \tag{28}$$

$$\forall l \in \mathbb{N} \forall n \geq K'(l) \ (|1 - \alpha_n| < 2^{-l}). \tag{29}$$

Then for any $x \in S$ and any natural number B' s.t. $B' \geq \|T^i x\|$ for all $i \leq K'(0)$ the sequence of the Cesàro means

$$A_n x := \frac{1}{n+1} \sum_{i=0}^{n} T^i x$$

is norm convergent and the following holds:

$$\forall l \in \mathbb{N}, g : \mathbb{N} \to \mathbb{N} \ \exists m \leq M'(l+1, g^{\mathsf{M}}, K, B) \forall i,j \in [m; m+g(m)] \ (\|A_i x - A_j x\| \leq 2^{-l}),$$

with $K(l) := K'(l + 2\lceil \log_2 B \rceil + 4)$, $B := 2B' + 1$ and M' defined as in Proposition 3.9.

Proof. Fix an arbitrary $x \in S$ and set $X_i := T^i x$, $\delta_n := 4B^2(\alpha_n^2 - 1)$. Next we show that the assumptions of Theorem 3.11 hold for (X_i), (δ_n), $K(l)$ and B. Obviously we have that $B \geq \|X_i\| + 1$ for all $i \leq K(0)$, since $2B' \geq \|T^i x\|$ for all $i \in \mathbb{N}$. Now, given any k,n and m, from $\|T^k x + T^k y\| \leq \alpha_k \|x + y\|$ for suitable x and y, we can infer that

$$\|X_{n+k} + X_{m+k}\|^2 \leq \|X_n + X_m\|^2 + (\alpha_k^2 - 1)\|X_n + X_m\|^2 \leq \|X_n + X_m\|^2 + (\alpha_k^2 - 1)4B^2.$$

Moreover, we have that (we can safely assume that $\alpha_n \geq 1$)

$$|\delta_n| = |4B^2(\alpha_n^2 - 1)| = 4B^2(\alpha_n - 1)(\alpha_n - 1 + 2) = 4B^2\left((\alpha_n - 1)^2 + 2(\alpha_n - 1)\right),$$

so using (29) we have that

$$|\delta_n| \leq 4B^2\left(2^{-2(l+2\lceil \log_2 B \rceil + 4)} + 2^{-(l+2\lceil \log_2 B \rceil + 3)}\right) \leq 4B^2 2^{-(l+2\lceil \log_2 B \rceil + 2)} \leq 2^{-l}.$$

Hence we obtain by Theorem 3.11 that

$$\forall l, g \exists m \leq M'(l+1, g^{\mathsf{M}}, K, B) \forall i,j \in [m; m+g(m)] \ (\|A_i - A_j\| \leq 2^{-l}),$$

with

$$A_n := \frac{1}{n} \sum_{i=1}^{n} X_i.$$

Finally, the claim follows from $X_i = T^i x$. □

Remark 3.13. Since the operator T is bounded by α_1 (it follows from (28) that $\forall x \in S\ \|Tx\| \leq \alpha_1 \|x\|$), we can easily compute a B' from two natural numbers a and b satisfying $a \geq \alpha_1$ and $b \geq \|x\|$. Therefore we can actually compute an $M''(l,g,K,a,b)$ s.t.

$$\forall l \in \mathbb{N}, g : \mathbb{N} \to \mathbb{N}\ \exists m \leq M''(l,g,K,a,b) \forall i,j \in [m; m+g(m)]\ (\|A_i x - A_j x\| \leq 2^{-l}).$$

The following corollary follows immediately from Corollary 3.12.

Corollary 3.14 (Finitary version of Theorem 2.1 in [129]). *Let S be a subset of a Hilbert space and $T : S \to S$ be a mapping satisfying*

$$\forall x,y \in S\ (\|Tx + Ty\| \leq \|x + y\|).$$

Then for any $x \in S$ and any natural number $b \geq \|x\|$ the sequence of the Cesàro means $A_n x := \frac{1}{n+1}\sum_{i=0}^{n} T^i x$ is norm convergent and the following holds:

$$\forall l \in \mathbb{N}, g : \mathbb{N} \to \mathbb{N}\ \exists m \leq M(l, g^{\mathsf{M}}, b)\ \forall i,j \in [m; m+g(m)]\ \ (\|A_i x - A_j x\| \leq 2^{-l}),$$

with M defined as follows:

$$\begin{aligned}
M(l,g,b) &:= (N(2l+7,g,b) + P(2l+7,g,b))b2^{2l+8} + 1,\\
P(l,g,b) &:= P_0(l, F(l,g,N(l,g,b),b),b),\\
F(l,g,n,b)(p) &:= p + n + \tilde{g}((n+p)b2^{l+1}),\ F(l,g,n,b) : \mathbb{N} \to \mathbb{N},\\
H(l,g,b)(n) &:= n + P_0(l, F(l,g,n,b)) + \tilde{g}((n + P_0(l, F(l,g,n,b),b))b2^{l+1}),\\
&\qquad\qquad H(l,g,b) : \mathbb{N} \to \mathbb{N},
\end{aligned}$$

where $P_0(l,f,b) := \tilde{f}^{b^2 2^l}(0)$, $\tilde{f}(n) := n + f(n)$ and, in this corollary, $f^{\mathsf{M}}(n) := \max_{i \leq n+1} f(i)$.

Note that (due to Lemma 3.18) the bound for m depends only on b, a bound for the norm of the parameter x, and not directly on the starting point itself.

3.5 Proof of Proposition 3.9

From now on, we assume that the assumptions of Proposition 3.9 hold and use the terms as they are defined in that proposition. Moreover, we will sometimes omit obvious dependency on trivial arguments, mainly the parameters K and B.
As a first step towards the proof, we define a specific primitive recursive 2^{-l} approximation of the square of the norm of the smallest convex combination of $X_n, X_{n+1}, \ldots, X_p$.

Remark 3.15. Note that while it is convenient to represent convex combinations of a fixed set of elements via finite tuples of rational numbers, not every tuple of rational numbers corresponds to such a convex combination. Moreover, we need a formal way to produce convex combinations of arbitrary length. Therefore we introduce for any tuple of rational numbers $\underline{s}$ a function $\tilde{\underline{s}} : \mathbb{N} \to \mathbb{N}$ which does have these properties and if $\underline{s}$ did correspond to a valid convex combination of desired length the result remains unchanged. Of course, it is not important to define $\tilde{\underline{s}}$ in a specific way (or at all).

Definition 3.16 (C). Let

$$C'(\underline{s}, l, n, p, X) := \left\| \sum_{i=0}^{p} \tilde{\underline{s}}(i) X_{n+i} \right\|^2$$

and

$$C(l,n,p) := C(l,n,p,X) := \min_{\underline{s} \in S_{p,l}} \{C'(\underline{s},l,n,p,X)\},$$

where

$$S_{p,l} := \left\{(s_0,\ldots,s_p) \,\middle|\, \sum_{i=0}^{p} s_i = 1 \wedge \forall i \in [0;p]\, \exists k_i \leq \left\lceil \frac{pB^2}{2^{-(l+1)}} \right\rceil \left(s_i = k_i \frac{2^{-(l+1)}}{pB^2}\right)\right\},$$

$$\widetilde{s_0,\ldots,s_m}(n) := \begin{cases} s_n & \text{if } n < m \wedge \ 0 \leq s_n \wedge s_n + \sum_{i=0}^{n-1} \tilde{\underline{s}}(i) \leq 1, \\ 0 & \text{if } n < m \wedge \neg(0 \leq s_n \wedge s_n + \sum_{i=0}^{n-1} \tilde{\underline{s}}(i) \leq 1), \\ s_m & \text{if } n = m \wedge \ 0 \leq s_m \wedge s_m + \sum_{i=0}^{m-1} \tilde{\underline{s}}(i) = 1, \\ 1 - \sum_{i=0}^{m-1} \tilde{\underline{s}}(i) & \text{if } n = m \wedge \neg(0 \leq s_m \wedge s_m + \sum_{i=0}^{m-1} \tilde{\underline{s}}(i) = 1), \\ 0 & \text{else.} \end{cases}$$

Lemma 3.17 (C approximates the smallest convex combination). *For any tuple of rational numbers $\underline{s}$ we have that*

$$\forall l,n,p,X\, \left(C'(\underline{s},l,n,p) + 2^{-l} \geq C(l,n,p)\right).$$

Proof. Given $\underline{s}$ choose $\underline{s}' \in S_{p,l}$ s.t. $|s'_i - \tilde{\underline{s}}(i)| \leq \frac{2^{-(l+1)}}{pB^2}$ for all $i \in [0;p]$. Then we have that

$$\left\| \sum_{i=0}^{p} \tilde{\underline{s}}(i) X_{n+i} - \sum_{i=0}^{p} s'_i X_{n+i} \right\| = \left\| \sum_{i=0}^{p} (\tilde{\underline{s}}(i) - s'_i) X_{n+i} \right\| \leq \frac{2^{-(l+1)}}{pB^2} pB = \frac{2^{-(l+1)}}{B},$$

and therefore also that $\left| \left\|\sum_{i=0}^{p} \tilde{\underline{s}}(i) X_{n+i}\right\| - \left\|\sum_{i=0}^{p} s'_i X_{n+i}\right\| \right| \leq \frac{2^{-(l+1)}}{B}$, so finally we get that

$$\left| \left\| \sum_{i=0}^{p} \tilde{\underline{s}}(i) X_{n+i} \right\|^2 - \left\| \sum_{i=0}^{p} s'_i X_{n+i} \right\|^2 \right| \leq \frac{2^{-(l+1)}}{B}\left(B + \frac{2^{-(l+1)}}{B}\right) \leq 2^{-l}.$$

□

We will also use that M majorizes itself (in the sense of Howard [51], see also [71]) and that N_0 and P_0 are the right witnesses for the two main assumptions needed in Wittmann's proof.

Lemma 3.18 (M is a majorant). *Each of the terms M,P,N,M_0,P_0,N_0 majorizes itself. In particular we have:*

$$\forall l \in \mathbb{N}, l' \geq l\, \forall h', h : \mathbb{N} \to \mathbb{N}\ (\forall n (h(n) \geq h'(n)) \to N_0(l', h^{\mathsf{M}}) \geq N_0(l, h'))$$

and

$$\forall l \in \mathbb{N}, g : \mathbb{N} \to \mathbb{N}\, \forall n \leq N(l, g^{\mathsf{M}}) \forall p \leq P(l, g^{\mathsf{M}})$$
$$(P(l, g^{\mathsf{M}}) \geq P_0(l, F(l, g^{\mathsf{M}}, n)) \ \wedge\ M(l, g^{\mathsf{M}}) \geq M_0(l, n, p)).$$

Lemma 3.19 (N_0 is correct).

$$\forall l \in \mathbb{N}, h : \mathbb{N} \to \mathbb{N}\, \exists n \leq N_0(l,h) \forall i,j \in [n; n+h(n)]\ \left(\|X_i\|^2 - \|X_j\|^2 \leq 2^{-l}\right).$$

Proof. W.l.o.g assume that $K = K^M$. The sequence $\|X_i\|^2$ is bounded by 0 and B^2 hence we can apply analogous version of Proposition 3.5 (actually the proof becomes even simpler, see [71], Proposition 2.26) to obtain that (recall that $P_0(l,u) = \tilde{u}^{B^2 2^l}(0)$):

$$\forall l, u \exists r \leq P_0(l,u) \forall i \leq u(r) \quad (\|X_i\|^2 + 2^{-l} \geq \|X_r\|^2). \tag{R}$$

Note that $N_0(l,h) = P_0(l+1,u) + K(l+1) \geq P_0(l,u)$ with $u(n) := n + K(l+1) + h^{\mathsf{M}}(n + K(l+1))$ so this implies that

$$\forall l, h \exists n \leq N_0(l,h) \quad \left(\big| \|X_{n+h(n)}\|^2 - \|X_n\|^2 \big| \leq 2^{-l} \right), \tag{N0}$$

since the following holds (here $N_0'(l,h,r) = r + K(l+1)$):

$$\begin{aligned}\forall l, h \exists r \leq P_0(l+1,u) \; (&\|X_{N_0'(l,h,r)+h(N_0'(l,h,r))}\|^2 \leq \|X_r\|^2 + 2^{-l-1} \wedge \\ &\|X_{N_0'(l,h,r)+h(N_0'(l,h,r))}\|^2 + 2^{-l-1} \geq \|X_r\|^2).\end{aligned}$$

The second inequality follows from (R) (for u as above) since

$$u(r) = r + K(l+1) + h^{\mathsf{M}}(r + K(l+1)) \geq N_0'(l,h,r) + h(N_0'(l,h,r)).$$

The first inequality follows from

$$\begin{aligned}\|X_{N_0'(l,h,r)+h(N_0'(l,h,r))}\|^2 &= \|X_{r+K(l+1)+h(N_0'(l,h,r))}\|^2 \\ &\leq \|X_r\|^2 + \frac{\delta_{K(l+1)+h(N_0'(l,h,r))}}{4} \leq \|X_r\|^2 + 2^{-l-1}.\end{aligned}$$

Note that for all $r \leq P_0(l,u)$ we have that $N_0'(l,h,r) \leq N_0'(l,h,P_0(l,u)) \leq N_0(l,h)$. Finally, given any h in the claim, we can define

$$h'(n) := \min\left\{ i \in [0; h(n)] \,\middle|\, \forall j \in [0; h(n)] \; \left(\big| \|X_{n+i}\|^2 - \|X_n\|^2 \big| \geq \big| \|X_{n+j}\|^2 - \|X_n\|^2 \big| \right) \right\}.$$

Now the claim follows from (N0) applied to h', the triangle inequality and the fact that we actually prove not only that $\|X_i\|^2 - \|X_j\|^2 \leq 2^{-l}$ but also $| \|X_i\|^2 - \|X_j\|^2 | \leq 2^{-l}$, and $N_0(l,h') \leq N_0(l,h^{\mathsf{M}})$, which follows from lemma 3.18 (we discuss a similar argument in the proof of Theorem 3.11 in more detail).

□

Lemma 3.20 (P_0 is correct).

$$\forall l, n \in \mathbb{N} \, \forall f : \mathbb{N} \to \mathbb{N} \, \exists p \leq P_0(l,f,B) \; \left(C(l,n,p) \leq C(l,n,f(p)) + 2^{-l} \right).$$

Proof. Fix an n and l arbitrarily. The sequence (a_i) defined by $a_i := C(l,n,i)$ is monotone, since for $i < j$ we have $\forall \underline{s} \in S_{i,l} \exists \underline{s}' \in S_{j,l} \; \widetilde{\underline{s}} = \widetilde{\underline{s}'}$ and therefore also $C(l,n,i) \geq C(l,n,j)$. Moreover (a_i) is a sequence in $[0, B^2]$, since $0 \leq C(l,n,i) \leq \|X_n\|^2 \leq B^2$ (note that $B \geq 1$) for any i. Hence by Proposition 3.5 we get

$$\forall k \in \mathbb{N}, g \in \mathbb{N}^{\mathbb{N}} \exists n \leq \tilde{g}^{B^2 \cdot 2^k}(0) \forall i,j \in [n; n+g(n)] \; \left(|a_i - a_j| < 2^{-k} \right),$$

which holds in particular also for $g = f$ and $k = l$ which implies that $P_0(l,f,B) = \tilde{f}^{B^2 2^l}(0) = \tilde{g}^{B^2 \cdot 2^k}(0)$. □

Next three lemmas give a quantitative analysis of the original proof in [129]. By l and g we always denote a natural number and a function $\mathbb{N} \to \mathbb{N}$ respectively.

Lemma 3.21 (The scalar product increase is bounded). *For any l and any g, consider $h : \mathbb{N} \to \mathbb{N}, h := H(l, g^{\mathsf{M}})$. Let n be a witness for Lemma 3.19, i.e.*

$$n \leq N(l,h) \ \wedge\ \forall i,j \in [n; n+h(n)] \ (i \leq j \to \|X_i\|^2 - \|X_j\|^2 \leq 2^{-l-1}). \qquad \text{(N)}$$

Moreover let $f := F(l, g^{\mathsf{M}}, n)$, p be a number smaller than $P_0(l,f)$ and $m := M_0(l,n,p)$. Then we have that

$$\langle X_{a+k}, X_{b+k} \rangle \leq \langle X_a, X_b \rangle + 2^{-l}$$

holds for all k, a, b s.t. $K^{\mathsf{M}}(l) \leq k \leq K^{\mathsf{M}}(l) + m + g^{\mathsf{M}}(m)$ and $n \leq a, b \leq n + p$.

Proof. We have

$$\|X_{a+k} + X_{b+k}\|^2 \leq \|X_a + X_b\|^2 + 2^{-l} \qquad (1)$$

since $k \geq K(l)$. Moreover we can infer

$$\|X_{a+k}\|^2 \geq \|X_a\|^2 - 2^{-l-1} \wedge \|X_{b+k}\|^2 \geq \|X_b\|^2 - 2^{-l-1} \qquad (2)$$

from (N), $a \geq n, b \geq n$, and

$$\begin{aligned} a+k, b+k &\leq n + p + m + g^{\mathsf{M}}(m) + K(l) \\ &= n + p + M_0(l,n,p) + g^{\mathsf{M}}(M_0(l,n,p)) + K(l) \\ &\leq n + H(l, g^{\mathsf{M}})(n) = n + h(n). \end{aligned}$$

Therefore

$$\begin{aligned} \langle X_{a+k}, X_{b+k} \rangle &= \frac{1}{2}(\|X_{a+k} + X_{b+k}\|^2 - \|X_{a+k}\|^2 - \|X_{b+k}\|^2) \\ &\leq \frac{1}{2}(\|X_a + X_b\|^2 + 2^{-l} - \|X_a\|^2 + 2^{-l-1} - \|X_b\|^2 + 2^{-l-1}) \\ &= \langle X_a, X_b \rangle + 2^{-l}. \end{aligned}$$

□

Analogously to Wittmann [129] we define

Definition 3.22 (Z). $Z(l,n,p,m) := \frac{1}{m+1} \sum_{k=K^{\mathsf{M}}(l)}^{K^{\mathsf{M}}(l)+m} \sum_{i=0}^{p} \widetilde{s}_i X_{n+k+i}$, with $\underline{s}$ corresponding to the tuple in the definition of $C(l,n,p)$ (see Definition 3.16 above).

Lemma 3.23 (Zs are close). *For any l and any g, consider $h := H(l, g^{\mathsf{M}})$. Let n be a witness for Lemma 3.19, i.e.*

$$n \leq N(l, g^{\mathsf{M}}) \ \wedge\ \forall i,j \in [n; n+h(n)] \ (i \leq j \to \|X_i\|^2 - \|X_j\|^2 \leq 2^{-l-1}). \qquad \text{(N)}$$

Moreover, let $m := M_0(l,n,p)$, $f := F(l, g^{\mathsf{M}}, n)$ and p be a witness for Lemma 3.20, i.e.

$$p \leq P_0(l,f) \ \wedge\ (C(l,n,p) \leq C(l,n,f(p)) + 2^{-l}), \qquad \text{(P)}$$

Then we have that $\|Z(l,n,p,m) - Z(l,n,p,m+g^{\mathsf{M}}(m))\|^2 \leq 2^{-l+4}$.

Proof. Firstly, we will show that

$$\|\frac{1}{2}(Z(l,n,p,m)+Z(l,n,p,m+g(m)))\|^2+2^{-l+1}\geq C(l,n,p). \tag{1}$$

Since $\frac{1}{2}(Z(l,n,p,m)+Z(l,n,p,m+g(m)))$ is a convex combination of

$$X_{n+K(l)},\ldots,X_{n+K(l)+p+m+g(m)},$$

we obtain by Lemma 3.17 that

$$\left\|\frac{1}{2}(Z(l,n,p,m)+Z(l,n,p,m+g(m)))\right\|^2+2^{-l}\geq \\ C(l,n,n+K(l)+p+m+g^{\mathsf{M}}(m)).$$

Now, because of

$$\begin{aligned} f(p) &= p+n+K(l)+M_0(l,n,p)+g^{\mathsf{M}}(M_0(l,n,p)) \\ &= n+K(l)+p+m+g^{\mathsf{M}}(m), \end{aligned}$$

it follows from (P) that

$$C(l,n,n+K(l)+p+m+g^{\mathsf{M}}(m))\geq C(l,n,p)-2^{-l},$$

which concludes the proof of (1). Secondly, we will show that

$$\forall o\leq m+g(m)\ \left(\left\|Z(l,n,p,o)\right\|^2\leq C(l,n,p)+2^{-l}\right). \tag{2}$$

Let $\underline{s}$ be the tuple corresponding to the tuple in the definition of $C(l,n,p)$ (note that $\tilde{\underline{s}}=\underline{s}$). By Lemma 3.21 we have

$$\begin{aligned} \left\|\sum_{i=0}^{p}s_iX_{n+k+i}\right\|^2 &= \sum_{i,j=0}^{p}s_is_j\langle X_{n+k+i},X_{n+k+j}\rangle \\ &\leq \sum_{i,j=0}^{p}s_is_j\langle X_{n+i},X_{n+j}\rangle+\sum_{i,j=0}^{p}s_is_j2^{-l}=\left\|\sum_{i=0}^{p}s_iX_{n+i}\right\|^2+2^{-l}, \end{aligned}$$

for all $K(l)\leq k\leq K(l)+m+g^{\mathsf{M}}(m)$, since $n\leq n+i,n+j\leq n+p$. Together with the convexity of the square function (and the definition of Z) this implies (2).
Finally, the claim follows from (1) and (2) by the parallelogram identity:

$$\begin{aligned} &\left\|Z(l,n,p,m)-Z(l,n,p,\tilde{g}(m))\right\|^2= \\ &=2\left\|Z(l,n,p,m)\right\|^2+2\left\|Z(l,n,p,\tilde{g}(m))\right\|^2-\left\|Z(l,n,p,m)+Z(l,n,p,\tilde{g}(m))\right\|^2 \\ &\leq 4(C(l,n,p)+2^{-l})-4(C(l,n,p)-2^{-l+1})=2^{-l+2}+2^{-l+3}\leq 2^{-l+4}. \end{aligned}$$

□

Lemma 3.24 (Zs and As are close). *For any l and any g, consider $h:=H(l,g^{\mathsf{M}})$. Let n be a witness for Lemma 3.19, i.e.*

$$n\leq N(l,g^{\mathsf{M}})\ \wedge\ \forall i,j\in[n;n+h(n)]\ \left(i\leq j\rightarrow\|X_i\|^2-\|X_j\|^2\leq 2^{-l-1}\right). \tag{N}$$

Moreover let $f := F(l, g^{\mathsf{M}}, n)$*,* p *be a witness for Lemma 3.20, i.e.*

$$p \leq P_0(l,f) \wedge (C(l,n,p) \leq C(l,n,f(p)) + 2^{-l}), \tag{P}$$

and $m := M_0(l,n,p)$*,* $m' := m + g(m)$*. Then we have that*

$$\|A_{m+1} - Z(l,n,p,m)\| \leq \frac{1}{m+1}(2n+2p+2K(l))B + 2^{-l}$$

and

$$\|A_{m'+1} - Z(l,n,p,m')\| \leq \frac{1}{m'+1}(2n+2p+2K(l))B + 2^{-l}.$$

Proof. From the definition of Z we see that (note that $m, m' \geq p$):

$$(m+1)Z(l,n,p,m) - \sum_{i=n+p+K(l)}^{n+K(l)+m} X_i = \sum_{i=0}^{p-1} t_i X_{n+K(l)+i} + \sum_{i=l}^{p} r_i X_{n+K(l)+m+i},$$

for suitable $\underline{t}$ and $\underline{r}$ with $0 \leq t_i, r_i \leq 1$. Hence (note that $m, m' \geq K(l) + n + p$)

$$\begin{aligned}
(m+1)\|Z(l,n,p,m) - A_{m+1}\| &= \left\| \sum_{k=K(l)}^{K(l)+m} \sum_{i=0}^{p} \underset{\sim}{s}_i X_{n+k+i} - \sum_{i=1}^{m+1} X_i \right\| \\
&= \left\| \sum_{i=0}^{p-1} t_i X_{n+K(l)+i} + \sum_{i=1}^{p} r_i X_{n+K(l)+m+i} + \sum_{i=n+p+K(l)}^{n+K(l)+m} X_i - \sum_{i=1}^{m+1} X_i \right\| \\
&= \left\| \sum_{i=0}^{p-1} t_i X_{n+K(l)+i} + \sum_{i=1}^{p} r_i X_{n+K(l)+m+i} + \sum_{i=m+2}^{n+K(l)+m} X_i - \sum_{i=1}^{n+p+K(l)-1} X_i \right\| \\
&\leq \left\| \sum_{i=0}^{p-1} t_i X_{n+K(l)+i} - \sum_{i=1}^{n+p+K(l)-1} X_i \right\| + \left\| \sum_{i=1}^{p} r_i X_{n+K(l)+m+i} \right\| + \left\| \sum_{i=m+2}^{n+K(l)+m} X_i \right\| \\
&\leq (n+p+K(l)-1)B + pB + (n+K(l)-1)B \leq (2n+2p+2K(l))B.
\end{aligned}$$

Obviously, same holds for m'.

□

Now, Proposition 3.9 can be proved as follows.

Proof of Proposition 3.9. Fix arbitrary l and g. Set $h := H(l, g^{\mathsf{M}})$. By Lemma 3.19 we know there is an n s.t.

$$n \leq N(l, g^{\mathsf{M}}) \wedge \forall i,j \in [n; n+h(n)] \left(i \leq j \to \|X_i\|^2 - \|X_j\|^2 \leq 2^{-l-1}\right).$$

Let $f := F(l, g^{\mathsf{M}}, n)$. By Lemma 3.20 we know that there is a p s.t.

$$p \leq P_0(l,f) \wedge (C(l,n,p) \leq C(l,n,f(p)) + 2^{-l}).$$

Note that by Lemma 3.18 we have that $p \leq P(l, g^{\mathsf{M}})$. We set $m := M_0(l,n,p)$. By Lemma 3.18 we get that $m \leq M(l, g^{\mathsf{M}})$. Finally, it follows from lemmas 3.23 and 3.24

that

$$\begin{aligned}\|A_{m+1} - A_{m+g(m)+1}\| &\leq \|Z(l,n,p,m) - Z(l,n,p,m+g(m))\| \\ &\quad + 2\Big(\frac{1}{m+1}(2n+2p+2K(l))B + 2^{-l}\Big) \\ &\leq \sqrt{2^{-l+4}} + 2^{-l+1} + \frac{2(2n+2p+2K(l))B}{m+1} \\ &= \sqrt{2^{-l+4}} + 2^{-l+1} + \frac{2(2n+2p+2K(l))B}{(2n+2p+2K(l))B2^l + 1} \\ &< \sqrt{2^{-l+4}} + 2^{-l+1} + 2^{-l+1} \leq 2^{-\frac{l}{2}+3}.\end{aligned}$$

This proves

$$\forall l, g \exists m \leq M(l, g^{\mathsf{M}})\ (\|A_{m+1} - A_{m+g(m)+1}\| \leq 2^{-\frac{l}{2}+3}),$$

from which the claim follows immediately by the definition of M'. □

Interpreting Bolzano-Weierstraß

The paper ... presents a direct computational interpretation of the Bolzano-Weierstrass principle (BW), via a combination of the negative translation and Gödel's dialectica interpretation (in the form of Shoenfield's interpretation). It is know in reverse mathematics that BW is equivalent to arithmetical comprehension. The goal of the paper is to find the precise complexity of the functional needed for the interpretation, in particular, when limited induction is available and a single instance of the BW principle is used ...
By naively looking at the proof of BW, one might think that Π^0_2 comprehension is needed even when a single instance of BW is used in a proof. The clever analysis in the paper shows that in fact all one needs is Σ^0_1 comprehension plus weak König's lemma (and WKL can be easily dealt with via the monotone interpretation). In fact, the authors work essentially directly with Σ^0_1-WKL ...
The authors show that a single instance of BW ... raises the complexity of the realiser by one ... The fact that this jump is optimal is argued ... This exploits the connection between BW and arithmetical comprehension, and standard results of Parson's on the complexity of induction for Σ^0_n formulas. The monotone functional interpretation of (a single instance of) BW is also considered, resulting in much simpler realisers ...

– anonymous referee about [108], Nov 23, 2009.

Motivation

In the context of proof minning, so far, sequential compactness could be dealt with using an elimination procedure due to U. Kohlenbach ([65, 66, 64, 68]) which replaces (if the underlying context is elementary enough) applications of fixed instances of these principles by their arithmetical counterparts. However, more substantial applications (e.g. in the context of weak compactness arguments) require solving their functional interpretation explicitly (see [72, 76]. In this section, we provide an explicit solution of the (negative translation of the) Gödel functional interpretation as well as the monotone functional interpretation of the Bolzano-Weierstraß theorem BW for $[0,1]$ and other compact metric spaces. Moreover, we show that the given solution is of optimal complexity and we will use it to obtain optimal program and bound extraction theorems for proofs based on BW applied to fixed bounded sequences in $[0,1]$ given by a term whose only free variables are the parameters of the theorem to be proved.

As is known from reverse mathematics ([114]) BW for compact metric spaces can be proved using CA_{ar} and – already for $[0,1]$ – also implies CA_{ar}. This is independent of whether BW is stated to assert the existence of a cluster point or a convergent subsequence. Moreover, these results can be proved relative to a very weak base system RCA_0. In fact, already the special case of BW stating that every monotone sequence in $[0,1]$ is convergent implies CA_{ar} (much refined results in this direction can be found in [68]).

It follows from Spector's solution of the functional interpretation of classical analysis by his bar recursive functionals $\mathcal{T} + (\mathsf{BR})$ ([117]) that the functional interpretation of CA_{ar} and hence of BW can be solved in the fragment $\mathcal{T}_0 + \mathsf{BR}_{0,1}$, where only the primitive recursor R_0 for type 0 and the bar recursor $\mathsf{B}_{0,1}$ for the types $0,1$ are used (see [67]). However, for a faithful calibration of the contribution of (single instances of) BW and the extraction of realizers of optimal complexity level from proofs of $\forall,\exists$-statements based on uses of BW even a functional interpretation in $\mathcal{T}_0 + \mathsf{BR}_{0,1}$ is too crude. It is crucial to note that the solution of the functional interpretation of BW uses only minimal number of nested simple $\mathsf{B}_{0,1}$-applications. In fact, we will see that a single use of $\mathsf{B}_{0,1}$ plus a use of a weak 'binary' form of bar recursion (due to Howard) suffices. Together with results of Howard and Parson this can be used to show that over systems such as

$$\widehat{\mathsf{WE\text{-}PA}}^{\omega}\!\restriction + \mathsf{QF\text{-}AC} + \Sigma^0_{n+1}\text{-}\mathsf{IA}$$

(which has a functional interpretation in $\mathcal{T}_n$, see [104]) the contribution of a use of BW in the form

$$\forall n\,(\mathsf{BW}(\xi(n)) \to \exists m\, \varphi_{\mathsf{QF}}(n,m))$$

in a proof of a sentence

$$\forall n \exists m\, \varphi_{\mathsf{QF}}(n,m) \text{ (with quantifier-free } \varphi_{\mathsf{QF}})$$

at most increases the complexity of the extractable algorithm f s.t.

$$\forall n\, \varphi_{\mathsf{QF}}(n, f(n))$$

from $f \in T_n$ to $f \in T_{n+1}$.
We will also show that this increase in general is necessary, thereby establishing the optimality of our result.

So far the aforementioned elimination method (see [64, 68]) has not been developed for systems based on $\mathcal{T}_0$-functionals and stronger ones as above. At the same time, the approach in the present section does not seem to be fine enough to re-obtain the results based on the elimination method for systems containing only Kalmar elementary functionals.

The functional dramatically simplifies if we switch to a majorizing functional in the sense of W.A. Howard (and hence to a solution of the monotone functional interpretation of BW). In particular, the use of Howard's 'binary' bar recursion then disappears altogether.

In [71], U. Kohlenbach has argued that the solutions provided by monotone functional interpretation of principles P directly correspond to the 'finitary' versions of P as discussed in Tao's program of 'finitary analysis' (see [123]). Following [123], the discussion in [71] focuses on the monotone convergence principle PCM and the infinitary pigeonhole principle IPP:

$$\mathsf{IPP} \;:\; \forall n \in \mathbb{N}\, \forall f : \mathbb{N} \mapsto \{0,\dots,n\}\, \exists i \le n \forall k \in \mathbb{N}\, \exists m \ge k\ (f(m) = i).$$

Already for IPP, it is nontrivial to arrive at a 'correct' finitization (see [30] for a thorough discussion). However, IPP is just a special case of BW for the discrete spaces $\{0,1,2,\ldots,n\}$. Hence we take one more step towards investigating the role of functional interpretations in connection with the program of finitizing analytical principles. In this section we also treat the Bolzano-Weierstraß principle for the compact (w.r.t. the product metric) metric space $\Pi_{i\in\mathbb{N}}[-k_i,k_i]$ (for sequences (k_i) in $\mathbb{R}_+$) whose functional interpretation has the same complexity as the one for the case $[0,1]$. This is of relevance for the logical analysis of proofs that use the weak compactness of closed, bounded convex sets in Hilbert spaces which can be reduced to the sequential compactness of $\Pi_{i\in\mathbb{N}}[-k_i,k_i]$ (see [72, 76]). Though certain uses of weak compactness in strong convergence results can be eliminated, in general it is unavoidable to deal with quantitative forms of weak compactness as provided by functional interpretation (see [73] as well as [76] for discussions).

For simplicity, let us come back to the case of $[0,1]$ and the Bolzano-Weierstraß theorem in the form stating that every sequence (x_n) of rational numbers in $[0,1]$ has a cluster point. In order to obtain a solution of the functional interpretation of optimal complexity one has to start with an appropriate proof of this statement: one standard way is to select one of the subintervals $[0,\frac{1}{2}],[\frac{1}{2},1]$ that contains infinitely many elements of the sequence (x_n) (by IPP at least one of the two intervals has this property) and then to continue with that interval. In this way one gets a nested sequence $I_0\supset I_1\supset I_2,\ldots$ of intervals I_k of length 2^{-k} that converges to a cluster point. In order to decide whether an interval I_k contains infinitely many elements of (x_n) one needs Π^0_2-comprehension since

$$\forall m\exists n\geq m\,(x_n\in I_k)\in\Pi^0_2.$$

However, in order to get the existence of just **some** sequence $I_0\supset I_1\supset I_2,\ldots$ of intervals I_n with the above property (rather than deciding this property which would be necessary only for finding – say – a left-most sequence, i.e. for constructing the limit inferior of (x_n) which indeed is of strictly greater complexity, see [68]) one can use König's lemma for $0/1$-trees. Note though, that this is not a use of what is called weak König's lemma (WKL) in reverse mathematics (see [114]) since the tree predicate is not quantifier-free but is Π^0_2 (and so, in fact, is an instance of what we call Π^0_2-WKL). Nevertheless, using a single instance of Σ^0_1-comprehension (short: Σ^0_1-CA) one can reduce such a Π^0_2-formula to a Π^0_1-tree predicate (by absorbing the inner existential quantifier). Now WKL for Π^0_1-trees (i.e. Π^0_1-WKL) can easily be reduced to the usual WKL. In this way the use of Π^0_2-comprehension is replaced by a use of Σ^0_1-CA plus the use of WKL, where the latter is known not to contribute to the complexity of extractable bounds. In fact, rather than first reducing Π^0_2-WKL using Σ^0_1-CA to Π^0_1-WKL and subsequently to WKL, we work directly with Σ^0_1-WKL and reduce this via Σ^0_1-CA in one step to WKL. The functional interpretation of BW, therefore, essentially boils down to solving the functional interpretation of Σ^0_1-WKL.

4.1 Interpreting Bolzano-Weierstraß

In this section we will use bar recursion to interpret the Bolzano-Weierstraß theorem.

Definition 4.1. The Bolzano-Weierstraß Principle
Let x be a sequence in $\mathbb{P}:=\prod_{i\in\mathbb{N}}[-k_i,k_i]$ for a known sequence $(k_i)_{i\in\mathbb{N}}$ with k_i in $\mathbb{Q}^+$.

Let d^ω denote the standard product metric (as defined e.g. in [114]):

$$d^\omega(a,b) :=_{\mathbb{R}} \sum_{i=0}^{\infty} \frac{1}{2^i} \frac{|a_i - b_i|}{1+|a_i - b_i|}, \text{ for } a,b \in \mathbb{P}.$$

In the following we tacitly rely on our representation of real numbers by which sequences of real numbers are represented by objects $a^{1(0)}$ and sequences of sequences of real numbers by objects $x^{1(0)(0)}$ (and each such object is a representative of a unique such sequence). We define

$$\mathsf{BW}_{\mathbb{R}}^{\omega} \quad : \quad \forall x^{1(0)(0)} \in \mathbb{P}^{\mathbb{N}} \underbrace{\exists a^{10} \in \mathbb{P} \ \forall k^0 \exists l^0 \geq_0 k \ d^\omega(xl,a) \leq_{\mathbb{R}} 2^{-k}}_{\equiv:\mathsf{BW}_{\mathbb{R}}^{\omega}(x^{1(0)(0)})},$$

where by $x \in \mathbb{P}^{\mathbb{N}}$ we mean a sequence of elements of $\mathbb{P}$ – i.e. a sequence (x) of sequences $(xn \in \mathbb{P},\ n \in \mathbb{N})$ of real numbers in the corresponding intervals $((xn)i \in [-k_i,k_i], i \in \mathbb{N})$.[23]

A Simple Proof of BW based on Σ_1^0-WKL

To demonstrate the main idea of the proof we only treat $\mathsf{BW}_{\mathbb{R}}^{\omega}$ for sequences of rational numbers in the unit interval $[0,1]$ – denoted by $\mathsf{BW}_{\mathbb{Q}}$ – which obviously is implied by (and in fact equivalent to)

$$\mathsf{BW}'_{\mathbb{Q}} \quad : \quad \forall s^1 \underbrace{\exists a^1 \forall k^0 \exists l^0 \geq_0 k \ |\widetilde{\hat{a}(k+1)} -_{\mathbb{Q}} \widetilde{sl}\ | \leq_{\mathbb{Q}} \langle 2^{-(k+1)} \rangle}_{\equiv:\mathsf{BW}'_{\mathbb{Q}}(s^1)},$$

where $\tilde{n} := \min_{\mathbb{Q}}\{\langle 1 \rangle, \max_{\mathbb{Q}}\{\langle 0 \rangle, n\}\}$. Consider a tree representation of the unit interval $[0,1]$ which splits the unit interval at level n into 2^n intervals of length 2^{-n}. Note that we can define each node via the path from the root to this interval. This path can be represented by a binary sequence b, where the n-th element defines which branch to take.
We define a predicate $I(b^0,n^0,m^0)$, which tells us, whether the rational number r encoded by $m = \langle r \rangle$ belongs to an interval defined by such a finite binary sequence b of length $\geq n$, i.e. in an interval of length 2^{-n} given by b:

$$\begin{aligned} I(b^0,n^0,m^0) :\equiv \quad & n \leq_0 \mathrm{lh}(b) \wedge \left\langle \sum_{i=1}^{n} \frac{b_i}{2^i} \right\rangle \leq_{\mathbb{Q}} m \leq_{\mathbb{Q}} \left\langle \sum_{i=1}^{n} \frac{b_i}{2^i} + \frac{1}{2^n} \right\rangle \\ \Leftrightarrow \quad & n \leq \mathrm{lh}(b) \wedge \sum_{i=1}^{n} \frac{b_i}{2^i} \leq r \leq \sum_{i=1}^{n} \frac{b_i}{2^i} + \frac{1}{2^n}. \end{aligned}$$

We know that for a given finite binary sequence b and an infinite sequence of encodings of rational numbers s, there is a function $f_s^{1(0)}$, primitive recursive in b and s, such that:

$$f_s(b,k) =_0 0 \ \leftrightarrow \ \Big(k > \mathrm{lh}(b) \wedge I\big(b, \mathrm{lh}(b), \widetilde{sk}\big) \Big).$$

[23] One can easily construct effective transformations which assign to any x of the type $1(0)(0)$ (or $1(0)$) a unique $\tilde{x}$ in $\mathbb{P}^{\mathbb{N}}$ (or $\mathbb{P}$), see theorem 4.6.

Now, by Σ^0_1-CA(f_s) we obtain a function g_s, s.t. $\forall b^0 \left(g_s b =_0 0 \leftrightarrow \exists k^0 \left(f_s(b,k) =_0 0\right) \right)$. In other words we have for all b^0:

$$g_s(b) =_0 0 \leftrightarrow \exists k^0 >_0 \mathrm{lh}(b) \;\; I\big(b, \mathrm{lh}(b), \widetilde{sk}\big). \tag{+}$$

To show BinTree(g_s), consider any finite binary sequence b:

$$\begin{aligned} g_s(b) =_0 0 \wedge x \subseteq b &\to \exists k^0 >_0 \mathrm{lh}(b) \;\; I\big(b, \mathrm{lh}(b), \widetilde{sk}\big) \wedge \mathrm{lh}(x) \leq_0 \mathrm{lh}(b) \wedge x \subseteq b \\ &\to \exists k^0 >_0 \mathrm{lh}(x) \;\; I\big(b, \mathrm{lh}(x), \widetilde{sk}\big) \wedge x \subseteq b \\ &\to \exists k^0 >_0 \mathrm{lh}(x) \;\; I\big(x, \mathrm{lh}(x), \widetilde{sk}\big) \to g_s(x) =_0 0. \end{aligned}$$

To show

$$\forall k \exists x \big(\mathrm{lh}(x) =_0 k \wedge g_s(x) =_0 0\big) \tag{++}$$

just consider any given natural number k. By the definition of our tree, it splits the $[0,1]$ interval at any level, in particular on level k, completely. Therefore, we have: $I(b, \mathrm{lh}(b), \widetilde{sk})$ and $\mathrm{lh}(b) =_0 k$ for a suitable b. As we started with arbitrary k, this implies (++).

Now, we can apply WKL(g_s) to get:

$$\exists b^1 \big(\mathsf{BinFunc}(b) \wedge \forall k \; g_s(\overline{b}k) =_0 0\big). \tag{$*$}$$

Note that in $(*)$ (and from now on) b^1 is a binary function and g_s takes the encoding of the initial segment, $\langle b(0), \ldots, b(k-1)\rangle$, of this infinite sequence as its type 0 argument. Using (+) we can conclude that $(*)$ is equivalent to: $\exists b^1 \leq \mathbf{1} \; \forall n \; \exists k > n \;\; I(\overline{b}n, n, \widetilde{sk})$. This means that $\mathsf{BW}'_{\mathbb{Q}}(s)$ is satisfied by $\hat{a}$ where a is defined as:

$a(n^0) :=_{\mathbb{Q}} \left\langle \sum_{i=1}^{n+1} \frac{b(i-1)}{2^i} + \frac{1}{2^{n+2}} \right\rangle$ provided that $\widehat{a} =_1 a$.

It, therefore, remains to show that a represents a real number in the sense of Definition 1.13. W.l.o.g, at this point, we use $r, =, |\cdot|, \ldots$ directly instead of the proper syntactic form $\langle r\rangle, =_{\mathbb{Q}}, |\cdot|_{\mathbb{Q}}, \ldots$ to achieve better readability. To prove $a =_1 \hat{a} \in \mathbb{R}$, take any natural number n. We have:

$$\begin{aligned} |a(n) - a(n+1)| &= \left| \sum_{i=1}^{n+2} \frac{b(i-1)}{2^i} + \frac{1}{2^{n+3}} - \left(\sum_{i=1}^{n+1} \frac{b(i-1)}{2^i} + \frac{1}{2^{n+2}} \right) \right| \\ &= \left| \frac{b(n+1)}{2^{n+2}} + \frac{1}{2^{n+3}} - \frac{1}{2^{n+2}} \right| = 2^{-(n+3)} < 2^{-(n+1)}, \end{aligned}$$

which concludes the proof.

The only relevant difference for the general case (i.e. for sequences in $\mathbb{P}$) is the definition of f_s. If we wanted each node i at level n to define a subspace $\mathbb{P}^n_i \subseteq \mathbb{P}$ such that $\exists a^n_i \in \mathbb{P}^n_i \forall b \in \mathbb{P}^n_i d^\omega(a,b) \leq 2^{-n}$, then the number of children couldn't be bounded by a constant.

It turns out that it is simpler to define a representation of $\mathbb{P}$ by a binary tree, where any infinite path defines a single element of $\mathbb{P}$ and provide a function which returns the sufficient level to satisfy the condition above.

We define such a tree as follows. We start by splitting the first dimension into two halves, i.e. the two children represent the spaces $\mathbb{P}^1_0 = [-k_0, 0] \times \prod_{j=1}^{\infty}[-k_j, k_j]$ and

$\mathbb{P}_1^1 = [0, k_0] \times \prod_{j=1}^{\infty}[-k_j, k_j]$. Next two levels arise by first splitting the new intervals in the first dimension and then splitting the second dimension into two halves. At level $\frac{l(l+1)}{2}$ we create the next $l+1$ levels by splitting the new intervals for the first l dimensions and by splitting the original interval for the $(l+1)^{\text{th}}$ dimension. We define formally:

Definition 4.2. Let w be the primitive recursive function representing the number of times we split dimension d up to level n:

$$w(n^0, d^0) := \max_{\mathbb{N}}\{l : l > 0 \wedge 1 + \frac{(d+l)(d+l-1)}{2} + d \leq n \ \vee \ l = 0\}.$$

For an encoding of a finite binary sequence $b = \langle b_0, b_1, \ldots, b_{n-1}\rangle$ we define

- $D(b^0, d^0) := \langle b_{d(d+1)/2+d}, b_{(d+1)(d+2)/2+d}, \ldots, b_{(d+w(n,d)-1)(d+w(n,d))/2+d}\rangle$ (the code for the splittings of dimension d corresponding to the node defined by b – for $w(n,d) = 0$ we define D to be the empty sequence – using the Cantor pairing function we could also write $D(b^0, d^0) := \langle b_{\langle i,d\rangle} : i \in \{0, \ldots, w(n,d) - 1\}\rangle$),
- $\mathbb{P}_d^b := \begin{cases} [-k_d + \sum_{i=0}^{w(n,d)} \frac{D(b,d)(i)}{2^i} k_d, k_d - \sum_{i=0}^{w(n,d)} \frac{1-D(b,d)(i)}{2^i} k_d] & \text{if } d \leq w(n,0), \\ [-k_d, k_d] & \text{else.} \end{cases}$ (the partition of the dimension d relevant at the node defined by b),
- $\mathbb{P}^b := \prod_{i=0}^{\infty} \mathbb{P}_i^b$ (the subspace corresponding to the node defined by b).

For a sequence $s \subseteq \mathbb{P}$ we define f_s^1 as follows:

$$f_s(b,k) := \begin{cases} 0 & \text{if} \quad k > \mathrm{lh}(b) \ \wedge \ \forall i < \mathrm{lh}(b)\, b(i) \leq 1 \ \wedge \\ & \quad \bigwedge_{i=0}^{w(\mathrm{lh}(b),0)} (\widetilde{sk})ik \in_{\mathbb{Q}} \mathbb{P}_i^b, \\ 1 & \text{else.} \end{cases}$$

From now on our notation refers to this definition.

Lemma 4.3. *Define the functions*

$$\mathrm{lv}_d^1(n) :=_{\mathbb{N}} 1 + \frac{(\lceil \log_2(k_d(n+2))\rceil + n + 1)(\lceil \log_2(k_d(n+2))\rceil + n)}{2} + d,$$
$$\mathrm{lv}(n) :=_{\mathbb{N}} \max_{\mathbb{N}}\{\mathrm{lv}_d(n) : d \leq n+1\}.$$

Then the following holds for all finite binary sequences b and $n \in \mathbb{N}$:

$$\mathrm{lh}(b) \geq \mathrm{lv}(n) \to \forall x, y \in \mathbb{P}^b \ \left(d^\omega(x,y) \leq 2^{-n}\right).$$

For the specific x, s.t. x_d is the center of $\mathbb{P}_d^b$ for all dimensions d we have even:

$$A^{\mathrm{BW}}([b]) :=_{1(0)} \lambda d^0, n^0. \left\langle -k_d + \sum_{i=0}^{w(\mathrm{lv}(n),d)-1} \frac{D(\overline{[b]}(\mathrm{lv}(n)), d)(i)}{2^i} k_d + \frac{1}{2^{w(\mathrm{lv}(n),d)}} k_d \right\rangle,$$

$$\mathrm{lh}(b) \geq \mathrm{lv}(n) \to \forall y \in \mathbb{P}^b \ \left(d^\omega(A^{\mathrm{BW}}([b]), y) \leq 2^{-n-1}\right).$$

Moreover, we have

$$f_s(b,k) = 0 \wedge \mathrm{lh}(b) \geq \mathrm{lv}(n) \quad \to \quad d^\omega((\widetilde{sk}), A^{\mathrm{BW}}([b])) < 2^{-n}.$$

Proof. W.l.o.g let $l := \mathrm{lh}(b) = \mathrm{lv}(n)$. By definition we have $w(l,d) \geq \lceil \log_2(k_d(n+2)) \rceil + n + 1 - d$. This means $\lceil \log_2(k_d(n+2)) \rceil - w(l,d) \leq d - n - 1$.
So $(n+2)k_d 2^{-w(l,d)} \leq 2^{-n-1+d}$ and $2k_d - \sum_{i=0}^{w(l,d)} 2^{-i} k_d \leq \frac{2^{-n-1+d}}{n+2}$.
By definition of $\mathbb{P}_d^b$ we obtain for all $d \leq n+1 (\leq w(l,0))$: $|\mathbb{P}_d^b| \leq \frac{2^{-n-1+d}}{n+2}$,[24] which implies
$\sum_{d=0}^{n+1} 2^{-d} |\mathbb{P}_d^b| \leq 2^{-n-1}$ and $\sum_{d=0}^{\infty} 2^{-d} \frac{|\mathbb{P}_d^b|}{1+|\mathbb{P}_d^b|} \leq 2^{-n}$.
To show $d^\omega((\widetilde{sk}), A^{\mathsf{BW}}([b])) < 2^{-n}$ suppose $\bigwedge_{i=0}^{w(\mathrm{lh}(b),0)} (\widetilde{sk})ik \in \mathbb{P}_i^b$. This implies there is an $y \in \mathbb{P}^b$ s.t. $\forall d\, |y_d - (\widetilde{sk})_d| \leq 2^{-k}$. Therefore $d^\omega((\widetilde{sk}), A^{\mathsf{BW}}([b])) \leq 2^{-n-1} + 2^{-k}$ and since $k > \mathrm{lh}(b) \geq \mathrm{lv}(n) \geq n+2$ also $d^\omega((\widetilde{sk}), A^{\mathsf{BW}}([b])) < 2^{-n}$.

□

Furthermore, we need to show the following property of our tree representation of $\mathbb{P}$.

Lemma 4.4. *At any level $n \in \mathbb{N}$, the union of all spaces corresponding to the paths of length n is the whole space $\mathbb{P}$:*

$$\bigcup_{b \in \{b^0 : \mathrm{lh}(b) = n \wedge \bigwedge_{i=0}^{n} b(i) \leq_0 1\}} \mathbb{P}^b = \mathbb{P}.$$

Proof. Let I_d denote the set of indices within a given, arbitrary long binary sequences b used by D to generate the subsequence $D(b,d)$ ($I_d = \{(d+i)(d+i+1)/2 + d = \langle i, d \rangle : i \in \mathbb{N}\}$). Since the Cantor pairing function is bijective, it follows that for $d_1 \neq d_2$ the intersection $I_{d_1} \cap I_{d_2}$ is empty and we can choose the binary sequences for each dimension independently.
Therefore it suffices to show the following (we scale by $2k_d$ and shift by $\frac{1}{2}$):

$$\forall n \forall x \in [0,1] \exists b \left(\mathrm{lh}(b) = n \wedge x \in [\sum_{i=1}^{n} \frac{b(i-1)}{2^i}, 1 - \sum_{i=1}^{n} \frac{1 - b(i-1)}{2^i}]\right).$$

This holds for any n and x when we choose b as the following binary sequence:

$$b(i) := \begin{cases} 1 & \text{if } x \geq \sum_{j=1}^{i} \frac{b(j-1)}{2^j} + \frac{1}{2^{i+1}}, \\ 0 & \text{else.} \end{cases}$$

□

Functional Interpretation of $\mathsf{BW}_{\mathbb{R}}^{\omega}$

From now on we consider s to be an infinite sequence of points in $\mathbb{P}$, and f_s to be the characteristic function of the corresponding tree (as defined in 4.2 above). From section 1.16, we know that using an appropriate formula φ^{WKL} we can write WKL as

$$\forall (h^1) \mathsf{WKL}_\Delta(h) \equiv \forall (h^1) \exists b^1 \forall k\ \varphi^{\mathsf{WKL}}(h,b,k) \equiv \forall (h^1) \exists b^1 \forall k\ \widehat{\left(\widehat{h}\right)_{g(h)}}(\bar{b}k) =_0 0,$$

[24] By $|[a,b]|$, $a,b \in \mathbb{Q}$ we mean the length of the rational interval $[a,b]$.

where φ^{WKL} is quantifier-free.
We introduce the following notations for Σ^0_1-CA:

$$\begin{aligned}
&\exists g^1 \forall x^0 \varphi^{\mathsf{CA}(f)}_{\Sigma^0_1}(x, gx) \\
&\qquad \equiv \exists g^1 \forall x^0 (gx =_0 0 \leftrightarrow \exists z^0\, f(x,z) =_0 0) \\
\Leftrightarrow\; &\exists g^1 \forall x^0 \forall z^0_2 \exists z^0_1 \varphi^{\mathsf{CA}(f)}(x, gx, z_1, z_2) \\
&\qquad \equiv\; \exists g^1 \forall x^0 \forall z^0_2 \exists z^0_1 ((gx =_0 0 \rightarrow f(x,z_1) =_0 0) \wedge (gx =_0 0 \leftarrow f(x,z_2) =_0 0)),
\end{aligned}$$

where $\varphi^{\mathsf{CA}(f)}$ is a quantifier-free formula.
The essential step in the proof above is the following implication:

$$(\Sigma^0_1\text{-}\mathsf{CA}(f) \wedge \mathsf{WKL}) \quad \rightarrow \quad \left(\exists g, b \forall x, k \;(\; \varphi^{\mathsf{CA}(f)}_{\Sigma^0_1}(x, gx) \wedge \varphi^{\mathsf{WKL}}(g, b, k)\;)\right), \qquad (+)$$

since its conclusion is essentially the same as $\mathsf{WKL}(\psi_f)$, where $\psi_f(k^0) \leftrightarrow \exists n^0 f(k,n) =_0 0$ and k is the variable which is bound by the last for-all quantifier in WKL. Moreover, for f_s as defined above it actually directly implies BW, whereas considered as a schema for arbitrary f^1 it corresponds to Σ^0_1-WKL.
The Sh-interpretation of (+) using this representation and applying QF-AC is as follows:

$$\begin{aligned}
&\exists G, Z_1, B, H', X', Z'_2, K' \;\; \forall B', Z'_1, G', X, Z_2, K \\
&\quad \Big(\Big(\varphi^{\mathsf{CA}(f)}(X'(G'X'Z'_2)(Z'_1X'Z'_2), G'X'Z'_2(X'(G'X'Z'_2)(Z'_1X'Z'_2)), Z'_1X'Z'_2(X'(G'X'Z'_2)(Z'_1X'Z'_2))(Z'_2(G'X'Z'_2)(Z'_1X'Z'_2)), \\
&\qquad\qquad Z'_2(G'X'Z'_2)(Z'_1X'Z'_2)) \wedge \varphi^{\mathsf{WKL}}(H', B'H'K', K'(B'H'K'))\Big) \rightarrow \\
&\quad \Big(\varphi^{\mathsf{CA}(f)}(XGZ_1B, G(XGZ_1B), Z_1(XGZ_1B)(Z_2GZ_1B), Z_2GZ_1B) \wedge \varphi^{\mathsf{WKL}}(G, B, KGZ_1B)\Big)\Big)
\end{aligned}$$

where, again, each exists-variable (i.e. G, Z_1, B, H', X', Z'_2, and K') may depend on any for-all-variable (i.e. B', Z'_1, G', X, Z_2, and K). E.g. by G we mean in fact $(GB'Z'_1G'XZ_2K)$. This interpretation yields the following functional equations:

$$\begin{aligned}
X'(G'X'Z'_2)(Z'_1X'Z'_2) &= XGZ_1B, & H' &= G, \quad (1{,}5) \\
G'X'Z'_2(X'(G'X'Z'_2)(Z'_1X'Z'_2)) &= G(XGZ_1B), & B'H'K' &= B, \quad (2{,}6) \\
Z'_1X'Z'_2(X'(G'X'Z'_2)(Z'_1X'Z'_2))(Z'_2(G'X'Z'_2)(Z'_1X'Z'_2)) &= Z_1(XGZ_1B)(Z_2GZ_1B), & K'(B'H'K') &= KGZ_1B, \\
& & & \quad (3{,}7) \\
Z'_2(G'X'Z'_2)(Z'_1X'Z'_2) &= Z_2GZ_1B. & & \quad (4)
\end{aligned}$$

We use a very similar approach to the one used by Gerhardy in [33] to solve such equations for finite DNS. First, we conclude from (5) and (6) that $B = B'GK'$ and from (1) and (2) that $G = G'X'Z'_2$. Using (6), we can set K' to $\lambda b.KGZ_1b$ according to (7). This is not that trivial for X' and Z'_2. However, as pointed out by Gerhardy in [33], in the presence of the λg and λz_1, which as we know will stand for the input of G and Z_1, the objects X' and Z'_2 become well definable terms:

$$t_{X'} := \lambda g, z_1.Xgz_1(B'g(\lambda b.Kgz_1b)), \quad t_{Z'_2} := \lambda g, z_1.Z_2gz_1(B'g(\lambda b.Kgz_1b)).$$

This makes the rest of our terms we need well defined. This is easy to see since for each term all dependencies are only on the terms defined above:

$$\begin{aligned}
t_{Z_1} &:= Z'_1 t_{X'} t_{Z'_2}, & t_{H'} &:= t_G, \\
t_G &:= G' t_{X'} t_{Z'_2}, & t_{K'} &:= \lambda b.K t_G t_{Z_1} b, \\
t_B &:= B' t_G t_{K'}.
\end{aligned}$$

We have found the realizing terms for the Shoenfield interpretation $(+)^{Sh}$ of $(+)$ for any G', Z'_1 and B'. To finally obtain the Shoenfield interpretation of $(\Sigma^0_1\text{-WKL}(\varphi))$ we just need to define these three functionals in such a way that the assumptions φ^{WKL} and $\varphi^{\mathsf{CA}(f)}$ are always true.
For $\varphi^{\mathsf{CA}(f)}$, as we know from the functional interpretation of Σ^0_1-CA (see section 1.15), we get:

$$G'^3 = t_h, \quad Z_1'^3 = t_z,$$

where t_h and t_z are defined as in the Sh-interpretation of Σ^0_1-CA (see corollary 1.60).
For B', from the interpretation of WKL, we know the following equality holds:

$$\underbrace{\overline{B'H'K'}(K'(B'H'K'))}_{\text{as above}} \quad = \quad \underbrace{\overline{B}(AB)}_{\text{as in section 1.16}} \quad .$$

We use the same notation as we used to define B in section 1.16 and define:

$$B' := \lambda h.\lambda A.\big[F_h(K_A(\varnothing))\big],$$

where F_h and K_A are defined as in the Sh-interpretation of WKL (see Theorem 1.81).
The terms defined above, using these definitions for G' and B', then satisfy the Shoenfield interpretation of the conclusion of $(+)$:

$$\forall X, Z_2, K\ (\ \varphi^{\mathsf{CA}(f)}(Xt_Gt_{Z_1}t_B, t_G(Xt_Gt_{Z_1}t_B), t_{Z_1}(Xt_Gt_{Z_1}t_B)(Z_2t_Gt_{Z_1}t_B), Z_2t_Gt_{Z_1}t_B) \wedge \\ \varphi^{\mathsf{WKL}}(t_G, t_B, Kt_Gt_{Z_1}t_B)\ ).$$

Using that $\forall X^{0(10)(1)}, Z^{0(10)(1)}, a^0, b^0\ t_zXZab =_0 t_zXZa0 = t_Za0$ we conclude:

Lemma 4.5.
The principle (which essentially represents[25] Σ^0_1-WKL(φ)*):*

$$\exists g^1, b^1 \forall x^0, k^0\ (\varphi^{\mathsf{CA}(f)}_{\Sigma^0_1}(x, gx) \wedge \varphi^{\mathsf{WKL}}(g, b, k)),$$

is Sh-interpreted by ($\tau = 01(10)1$ [26]):

$$\forall X^\tau, Z^\tau, K^\tau\ \Big(\ \varphi^{\mathsf{CA}(f)}(Xt_Gt_Zt_B, t_G(Xt_Gt_Zt_B), t_Z(Xt_Gt_Zt_B)0, Zt_Gt_Zt_B) \wedge \\ \varphi^{\mathsf{WKL}}(t_G, t_B, Kt_Gt_Zt_B)\ \Big),$$

where

$$t_B :=_1 B^{\mathsf{WKL}}(\lambda b.Kt_Gt_Zb, t_G), \qquad t'_X :=_{0(10)1} \lambda g^1, z^{10}.Xgz(B^{\mathsf{WKL}}(\lambda b.Kgzb, g)),$$
$$t_Z :=_{10} t_z t'_X t'_Z, \qquad t'_Z :=_{0(10)1} \lambda g^1, z^{10}.Zgz(B^{\mathsf{WKL}}(\lambda b.Kgzb, g)),$$
$$t_G :=_1 t_h t'_X t'_Z,$$

The remaining terms are defined as in previous sections.[27]

[25] It trivially implies Σ^0_1-WKL(φ) (provably in $\widehat{\text{WE-PA}}^\omega\!\restriction$). Let us note, however, that the actual computation of the witnesses for Σ^0_1-WKL(φ) still involves some highly non-trivial technical work.
[26] Or, in a more illustrative notation: $\tau = (\mathbb{N} \to \mathbb{N}) \to (\mathbb{N} \to \mathbb{N} \to \mathbb{N}) \to (\mathbb{N} \to \mathbb{N}) \to \mathbb{N}$.
[27] B^{WKL} in Theorem 1.81, and t_z, t_h in Theorem 1.60.

Theorem 4.6. *The Bolzano-Weierstraß principle* $\mathsf{BW}^{\omega}_{\mathbb{R}}$ *for an infinite sequence s of elements in* $\mathbb{P}$ *(let* $\rho = (10)0$ *and* $\sigma = 0(10)1$, $\widetilde{x^{1(0)}} := \lambda i^0.\min_{\mathbb{R}}(-k_i, \max_{\mathbb{R}}(k_i, xi))$, *recall* $\mathbb{P} = \prod_{i\in\mathbb{N}}[-k_i, k_i]$*):*

$$\forall s^\rho \exists a^{1(0)} \forall m^0 \exists l^0 >_0 m \;\; d^\omega(\widetilde{sl}, \tilde{a}) <_{\mathbb{R}} 2^{-m}.$$

is Sh-interpreted[28] *by:*

$$\forall s^\rho, M^\sigma \exists L^1, a^{1(0)} \underbrace{\Big(\; L(MLa) >_0 MLa \;\wedge\; d^\omega(s(\widetilde{L(MLa)}), \tilde{a}) <_{\mathbb{R}} 2^{-MLa} \;\Big)}_{\mathsf{BW}_{Sh}(a,MLa,L(MLa),s):=}$$

where L and a are realized by the terms t^3_L *and* t^3_A *(we use the notation from Definition 4.2 and Lemma 4.3):*

$$\begin{aligned} t_L(s^\rho, M^\sigma) &:=_1 \lambda n^0.t_Z(\overline{t_B}(\mathrm{lv}(n)))0, \\ t_A(s^\rho, M^\sigma) &:=_{1(0)} A^{\mathsf{BW}}(t_B) \\ &= \lambda d^0, n^0. \Big\langle -k_d + \sum_{i=0}^{w(\mathrm{lv}(n),d)-1} \frac{D(\overline{t_B}(\mathrm{lv}(n)), d)(i)}{2^i} k_d + \frac{1}{2^{w(\mathrm{lv}(n),d)}} k_d \Big\rangle, \end{aligned}$$

Here t_B *is defined as above, i.e. (*B^{WKL} *is defined in Theorem 1.81):*

$$t_B =_1 B^{\mathsf{WKL}}(M' t_G t_Z, t_G),$$

where as before (with $f := f_s$, $z^- :\equiv_1 \lambda n.z^{10}n0$, $g^+ :\equiv_{10} \lambda a, b\,.g^1 a$ *and* $g\restriction_{f_s} :\equiv \lambda n.\overleftrightarrow{f_s(n, gn)}$*):*

$$t_Z =_{10} \Big(\Phi_0 \big(\lambda g^1.X(g\restriction_{f_s})g^+(B^{\mathsf{WKL}}(M'(g\restriction_{f_s})g^+, g\restriction_{f_s}))\big) u^{1(1(0))}_{(\lambda g^1.Z(g\restriction_{f_s})g^+(B^{\mathsf{WKL}}(M'(g\restriction_{f_s})g^+, g\restriction_{f_s}))), f_s} 0^0 0^1 \Big)^+$$

$$t_G =_1 ((t_Z)^-)\restriction_{f_s}.$$

Here t_B *and* t_Z *are shortcuts for* $t_B XZM'$ *and* $t_Z XZM'$ *with fixed X, Z:*

$$X :=_\tau \lambda g^1, z^{10}, b^1\,.\,B(M'gzb, g, z^-, b), \quad Z :=_\tau \lambda g^1, z^{10}, b^1\,.\,N(M'gzb, g, z^-, b), \;\; ,$$

where $(M')^\tau$ *(recall* $\tau = 01(10)1$ *) is defined for any given* M^σ *similarly as X and Z as follows:*

$$M'(g^1, z^{10}, b^1) :=_0 \mathrm{lv}\bigg(M \underbrace{\Big(\lambda n^0.z^-(\overline{b}(\mathrm{lv}(n)))\Big)}_{\sim t_L} \underbrace{\Big(A^{\mathsf{BW}}(b)\Big)}_{\sim t_A} \bigg). \tag{1}$$

The terms B, X_n[29] *and N are primitive recursive, though not trivial, case distinctions:*

$$B(m^0, g^1, z^1, b^1) :=_0 \begin{cases} \min_0 \left\{ x^0 \;\middle|\; \begin{array}{l} \mathrm{lh}(x) \le_0 m \wedge x \in \{0,1\}^{\mathrm{lh}(x)} \wedge \\ \neg\varphi^{\mathsf{CA}}(x, gx, zx, Nmgzb) \end{array} \right\} & \textit{if it exists,} \\ X_n(m, g, b) & \textit{else,} \end{cases}$$

$$X_n(m^0, g^1, b^1) :=_0 \begin{cases} \overline{b}m & \textit{if } g(\overline{b}m) =_0 0, \\ \min_0\{x^0 \mid \mathrm{lh}(x) = m \wedge f_s(x, m+1) =_0 0\} & \textit{else,} \end{cases}$$

[28] The quantifier in $<_{\mathbb{R}}$ is irrelevant, see the remark after the theorem.
[29] The existence of the required x follows from Lemma 4.4.

$$N(m^0, g^1, z^1, b^1) :=_0 \begin{cases} m+1 & \text{if } f_s(X_n(m,g,b), m+1) =_0 0, \\ z(X_n mgb) & \text{else.} \end{cases}$$

Proof. Unwinding φ^{CA} and φ^{WKL} we get by lemma 4.5:

$$\forall K^\tau, X^\tau, Z^\tau \ ((t_G XZK(X(t_G XZK)(t_Z XZK)(t_B XZK))) =_0 0 \to \\ f_s((X(t_G XZK)(t_Z XZK)(t_B XZK)), t_Z XZK(X(t_G XZK)(t_Z XZK)(t_B XZK))0) =_0 0) \quad (2)$$

and

$$\forall K^\tau, X^\tau, Z^\tau \ ((t_G XZK(X(t_G XZK)(t_Z XZK)(t_B XZK))) =_0 0 \leftarrow \\ f_s(X(t_G XZK)(t_Z XZK)(t_B XZK), Z(t_G XZK)(t_Z XZK)(t_B XZK)) =_0 0) \quad (3)$$

and

$$\forall K^\tau, X^\tau, Z^\tau \ \overbrace{(\widehat{t_G XZK})_{g(t_G XZK)} \ (\overline{t_B XZK}}(K(t_G XZK)(t_Z XZK)(t_B XZK))) =_0 0. \quad (4)$$

Fix an arbitrary M^σ.
We set X, Z and $K :=_\tau M'$ – see (1) – as in the theorem. We will use the following abbreviations:

$$\begin{aligned} x_0 &:\equiv_0 X(t_G XZK)(t_Z XZK)(t_B XZK), & \gamma &:\equiv_1 t_G XZK, \\ z_0 &:\equiv_0 Z(t_G XZK)(t_Z XZK)(t_B XZK), & z &:\equiv_1 \lambda n^0 . t_Z XZKn0, \\ k_0 &:\equiv_0 K(t_G XZK)(t_Z XZK)(t_B XZK), & b &:\equiv_1 t_B XZK. \end{aligned}$$

Note that by (2) the equality $\gamma(x_0) =_0 0$ implies $f_s(x_0, zx_0) =_0 0$ and thereby $\mathsf{BinFunc}([x_0])$. We will not be able to show $\mathsf{BinTree}(\gamma)$ but fortunately we need only to show:

$$\forall x^0 \ \Big((x \subseteq x_0 \ \wedge \ \gamma(x_0) = 0) \to \gamma(x) =_0 0 \Big) \quad (5)$$

and

$$\mathrm{lh}(x_0) =_0 k_0 \ \wedge \ \gamma(x_0) =_0 0. \quad (6)$$

Note that if x_0 was not equal to $X_n(k_0, \gamma, b)$ then

$$\neg\varphi^{\mathsf{CA}}(x_0, \gamma x_0, zx_0, \underbrace{N(k_0, \gamma, z, b)}_{=z_0})$$

would hold, which is a contradiction to (2) or a contradiction to (3). So we can assume that $x_0 = X_n(k_0, \gamma, b)$. Similarly, we have that

$$(\gamma(q) =_0 0 \to f_s(q, z(q)) =_0 0) \ \wedge \ (\gamma(q) =_0 0 \leftarrow f_s(q, z_0) =_0 0). \quad (7)$$

Suppose (7) would not hold for some q' with $\mathrm{lh}(q') \leq k_0$, then x_0 is equal to such a q' by the definition of X and we get a contradiction to (2) $\wedge$ (3) again.

- To prove (5) suppose:

$$x \subseteq x_0 \quad \wedge \quad (8a)$$

$$\gamma(x_0) = 0 \quad (8b)$$

holds for some x. Together with (8b) we obtain from (7):

$$f_s(x_0, z(x_0)) =_0 0,$$

since $\mathrm{lh}(x_0) \leq k_0$ (by the definition of X). We follow the definition of Z. We see that either z_0 directly equals $z(X_n(k_0, \gamma, b))$ or we have $f_s(X_n(k_0, \gamma, b), k_0 + 1) = 0$ and it equals $k_0 + 1$. This means that in both cases we obtain (using that $x_0 = X_n(k_0, \gamma, b)$)

$$f_s(x_0, z_0) =_0 0.$$

By the definition of f_s, see also section 4.1, and (8a) this implies

$$f_s(x, z_0) =_0 0.$$

From (8a) we get $\mathrm{lh}(x) \leq k_0$ and by (7) we obtain

$$\gamma(x) = 0,$$

which concludes the proof of (5).

- Recall that $x_0 = X_n(k_0, \gamma, b)$. This proves the first part of (6):

  $$\mathrm{lh}(x_0) =_0 k_0.$$

  Now we follow the definition of X_n. Either $\gamma(\bar{b}(k_0)) = 0$ and therefore $x_0 = \bar{b}(k_0)$ and we obtain (6) immediately, or we have that:

  $$f_s(X_n(k_0, \gamma, b), k_0 + 1) = 0.$$

  In that case, we can infer that

  $$z_0 = N(k_0, \gamma, z, b) = k_0 + 1,$$

  and we get

  $$f_s(x_0, z_0) = 0.$$

  Finally, applying (3) concludes the proof of (6).

This concludes the proofs of (5) and (6).
To show $\gamma(\bar{b}(k_0)) = 0$ assume towards contradiction that

$$\gamma(\bar{b}(k_0)) \neq 0.$$

If so, then by definition 1.80 we have also that

$$\widehat{\gamma}\,(\bar{b}(k_0)) \neq 0.$$

By (6) we know that $\gamma(x_0) = 0$. Using (5) we know that also $\widehat{\gamma}\,(x_0) = 0$ and since $\mathrm{lh}(x_0) = k_0 = \mathrm{lh}(\bar{b}(k_0))$ we have that

$$\widehat{\gamma}_{g\gamma}(\bar{b}(k_0)) \neq 0.$$

By definition 1.80 this is a contradiction to (4) and we obtain (recall that we started with an arbitrary M):

$$\forall M^\sigma\ \gamma(\overline{b}(k_0)) =_0 0.$$

This implies that $x_0 = \overline{b}(k_0)$ (by the definition of X_n) and therefore it follows by (2) that:

$$\forall M^\tau\ \ f_s(\ \overline{b}(k_0), z(\overline{b}k_0)\ ) =_0 0.$$

Using the terms t_L and t_A in the short notation (i.e. t_L instead of $t_L sM$, t_B instead of $t_B XZM'$ and similarly for t_A and t_Z) this becomes

$$\forall M^\sigma\ \ f_s(\ \overline{t_B}(\mathrm{lv}(Mt_Lt_A)),\ t_Z(\overline{t_B}(\mathrm{lv}(Mt_Lt_A)))0\ ) =_0 0.$$

This implies by Lemma 4.3 (note that $t_L(Mt_Lt_A) \equiv t_Z(\overline{t_B}(\mathrm{lv}(Mt_Lt_A)))0$):

$$\forall M^\sigma\ \Big(\ t_L(Mt_Lt_A) > Mt_Lt_A\ \wedge\ d^\omega(\widetilde{t_A}, s(\widetilde{t_L(Mt_Lt_A)}))) <_{\mathbb{R}} 2^{-Mt_Lt_A}\Big).$$

Finally observe that for all n^0 we have $(t_A(s,M))n =_{\mathbb{R}} (\widetilde{t_A(s,M)})n$ and that

$$\begin{aligned} t_Z &=_{10} t_z t'_X t'_Z = \big(t_g(t_f t'_X)(t_f t'_Z)\big)^+ \\ &= \lambda a,b\ .\ \underbrace{t_g(\lambda g^1.X(g{\restriction}_{f_s})g^+(B^{\mathsf{WKL}}(M'(g{\restriction}_{f_s})g^+, g{\restriction}_{f_s})))(\lambda g^1.Z(g{\restriction}_{f_s})g^+(B^{\mathsf{WKL}}(M'(g{\restriction}_{f_s})g^+, g{\restriction}_{f_s})))a}_{=\Phi_0\left(\lambda g^1.X(g{\restriction}_{f_s})g^+\left(B^{\mathsf{WKL}}(M'(g{\restriction}_{f_s})g^+, g{\restriction}_{f_s})\right)\right)u^{1(1(0))}_{\left(\lambda g^1.Z(g{\restriction}_{f_s})g^+\left(B^{\mathsf{WKL}}(M'(g{\restriction}_{f_s})g^+, g{\restriction}_{f_s})\right)\right), f_s}{}^{0^0 0^1 a}}, \\ t_G &=_1 t_h t'_X t'_Z = \big(t_g(t_f t'_X)(t_f t'_Z)\big){\restriction}_{f_s}. \end{aligned}$$

□

Remark. From the proof we actually see how to realize the hidden quantifier in $<_{\mathbb{R}}$. Namely before using Lemma 4.3 we can simply apply the definition of f_s and obtain the following equivalent universal formula (for $\rho = (10)0$ and $\sigma = 0(10)1$ and writing t_L instead of $t_L sM$ and t_A instead of $t_A sM$):

$$\forall s^\rho, M^\sigma \big(t_{L(Mt_Lt_A)} >_0 Mt_Lt_A \wedge \bigwedge_{d=0}^{w(\mathrm{lv}(Mt_Lt_A))} (s(\widetilde{t_{L(Mt_Lt_A)}}))d(t_L(Mt_Lt_A)) \in_{\mathbb{Q}} \mathbb{P}^{\overline{t_B}(\mathrm{lv}(Mt_Lt_A))}.$$

Remark. One can also extend Theorem 4.6 to cover the case where the sequence (k_d) is a sequence in $\mathbb{R}_+$ rather than $\mathbb{Q}_+$ provided that one adapts the definition of f_s in such a way that also the boundaries of the interval $[-k_d, k_d]$ are replaced by suitable rational approximations etc. Since the application of our analysis of the Bolzano-Weierstraß theorem given in [73] (referred to in the introduction) only uses the monotone functional interpretation given below (Theorem 4.10) as will be usually the case, we restrict ourselves in this thesis to treat the case with real k_d only in that context where things are particularly simple (and no approximation of the type mentioned above is needed).

Usually, the Bolzano-Weierstraß theorem is formulated to state the existence of a converging subsequence rather than the existence of a cluster point. The next theorem gives the solution for the Shoenfield interpretation of this formulation:

Theorem 4.7. *The version of the Bolzano-Weierstraß theorem stating the existence of a convergent subsequence is Sh-interpreted as follows (where t_L, t_A are as in theorem 4.6):*

$$\forall s, M \exists f, p(f(M f p + 1) > f(M f p) \wedge d^{\omega}(\tilde{p}, s(f(\widetilde{M f p} + 1)))) < 2^{-M f p})$$

Let M' be obtained from M by

$$M' l a := \begin{cases} l^{M(\lambda n . l^n 0) a} 0 & \text{if } (\forall n < M(\lambda n . l^n 0) a) \ (l^{n+1} 0 > l^n 0), \\ l^{\min_0 \{n : l^{n+1} 0 \le l^n 0\}} 0 & \text{else.} \end{cases}$$

Define the functionals

$$t_F s M := \lambda n . (t_L s M')^n 0 \quad \text{and} \quad t_P s M := t_A s M',$$

then the following holds:

$$\forall s, M \quad (\ t_F s M(M(t_F s M)(t_P s M) + 1) > t_F s M(M(t_F s M)(t_P s M)) \wedge$$
$$d^{\omega}(s(t_F s M(M(\widetilde{t_F s M})(t_P s M) + 1)), \widetilde{t_P s M}) < 2^{-M(t_F s M)(t_P s M)} \ ).$$

Proof. Consider any given s and M and define M' as in the theorem being proved. We denote $t_L s M'$, $t_A s M'$ and $M(\lambda n . (t_L s M')^n 0)(t_A s M')$ by l_0, a_0 and m_0. Unwinding the terms t_P and t_F in the statement of the theorem leads to:

$$\forall s, M \quad (\ l_0^{m_0+1} 0 > l_0^{m_0} 0 \wedge d^{\omega}(\tilde{a}_0, s(\widetilde{l_0^{m_0+1}} 0)) \le 2^{-m_0} \ ).$$

1. If $(\forall n < m_0) \ (l_0^{n+1} 0 > l_0^n 0)$ then the claim follows directly from theorem 4.6 applied to s and M' (i.e. $M' l_0 a_0$ becomes $l_0^{m_0} 0$) and its proof:

$$l_0(l_0^{m_0} 0) > l_0^{m_0} 0 \wedge d^{\omega}(s(\widetilde{l_0(l_0^{m_0} 0)}), \tilde{a}_0) <_{\mathbb{R}} 2^{-l_0^{m_0} 0} \qquad (4.6 \text{ for } M')$$

where the proof allows us to omit the final step from:

$$l_0(l_0^{m_0} 0) >_0 l_0^{m_0} 0 \wedge \bigwedge_{d=0}^{l_0^{m_0} 0} |(a_0)(l_0^{m_0} 0 + 1)d -_{\mathbb{Q}} s(\widetilde{l_0(l_0^{m_0} 0)})d(l_0^{m_0} 0 + 1)| \quad <_{\mathbb{Q}} \quad 2^{-(l_0^{m_0} 0 + 2)}.$$

To see that $l_0^{m_0} \ge m_0$ recall that $\forall n < m_0 \ l_0^{n+1} 0 > l_0^n 0$.

2. Otherwise we have $l_0^{i+1} 0 \le l_0^i 0$ for some $i < m_0$ and $M' l_0 a_0 = l_0^i 0$. However this is a contradiction to theorem 4.6 which implies

$$l_0^{i+1} 0 = l_0(l_0^i 0) = l_0(M' l_0 a_0) > M' l_0 a_0 = l_0^i 0.$$

$\square$

Remark. Analogously to the previous remark, the proof of Theorem 4.6 gives us a more strict
Sh-interpretation.

Applying the majorization results from section 1.16 we obtain a (much easier) solution to the **monotone** Shoenfield interpretation of $\mathsf{BW}^{\omega}_{\mathbb{R}}$ (which is all the information on $\mathsf{BW}^{\omega}_{\mathbb{R}}$ one needs to extract uniform bounds from proofs that use $\mathsf{BW}^{\omega}_{\mathbb{R}}$).
Instead of having to majorize explicitly the complicated construction of t_A we can rely on the fact that elements in compact intervals $[-K_d, K_d]$ with $K_d \in \mathbb{N}$ have a representations $x \leq_1 N_{K_d}$, where $N_m(k)$ is a fixed (non-decreasing) primitive recursive function in m, k which can be taken as (see [71], p.93)

$$N_m(k) := j(m2^{k+3} + 1, 2^{k+2} - 1) \text{ for the Cantor pairing function } j.$$

From the proof of Theorem 4.6 it follows that instead of $(A^{\mathsf{BW}}(b))(d)$ we can take any other representative of the same real number, in particular the representative $\leq N_{K_d}$ from the construction in [71](p.93) for $K_d \geq k_d$. As a result, we can replace $t_A(s, M)(d)$ by another representative that is – for all s, M – majorized by N_{K_d}. In fact, we may even allow k_d to be a real number where then in the definition of f_s the clause '$(\widetilde{sk})ik \in_{\mathbb{Q}} \mathbb{P}^{b}_{i}$' has to be replaced by '$(\widetilde{sk})i \in_{\mathbb{R}} \mathbb{P}^{b}_{i}$'. The resulting function f_s then no longer is computable but still trivially majorizable by the constant-1 function.

Remark. The shape of N_m above is due to the particular Cauchy representation used in [71]. If one uses the so-called signed-digit representation, it can be improved to the constant-$(2m + 3)$ function, see prop. 14 in [24].

Before we state the monotone functional interpretation of $\mathsf{BW}^{\omega}_{\mathbb{R}}$ we first need, however, a simple lemma:

Lemma 4.8. *Define (for* $\mathbb{N} \ni K_d \geq k_d$ *for all* $d \in \mathbb{N}$*)*

$$\mathrm{lv}^*(n) := (\max_{d \leq n+1} \{K_d, n\} + 2)^4.$$

Then lv^* maj lv.

Proof. observe that for $d \leq n + 1$ we have that

$$\mathrm{lv}_d(n) \leq n+2+(\lceil k_d(n+2)\rceil + n)^2 \leq (n+2) + (n+2)^2(\lceil k_d + 1\rceil)^2 \leq (n+2)^2(\lceil k_d + 2\rceil)^2.$$

Moreover lv^* is non-decreasing. □ To give the monotone version of Theorem 4.7, we need the following definition:

Definition 4.9. A majorant for Φ_0 is given by (see also [71] or [15]):

$$\Phi_0^* y u^{0(1(0))(0)} n x :=_1 \max\left((\Phi_0 y^{\mathsf{m}} u_x^{0(1(0))(0)})^{\mathsf{M}} n x, x^{\mathsf{M}} \right),$$

where

$$\begin{aligned} w^{\mathsf{M}}(n) &:=_\rho \max\{w(i) : i \leq_0 n\} \quad (\text{for } w^{\rho 0}, \rho \in \boldsymbol{T}), \\ y^{\mathsf{m}}(x) &:=_0 y(x^{\mathsf{M}}), \\ u_x(n, v) &:=_1 \max\{x^{\mathsf{M}}, v(unv)\}. \end{aligned}$$

Theorem 4.10. *Let* (k_d) *be a sequence of non-negative reals and* (K_d) *be a sequence of natural numbers with* $K_d \geq k_d$ *for all* $d \in \mathbb{N}$. *The realizing terms of the Sh-interpretation of the*

Bolzano-Weierstraß principle $\mathsf{BW}^{\omega}_{\mathbb{R}}$ *for an infinite sequence s of elements in* $\mathbb{P}$ *(defined by* (k_d)*, see also Theorem 4.6 above) can be majorized by terms* t^*_L *and* t^*_A*, i.e.* t^*_L *and* t^*_A *satisfy the monotone Sh-interpretation of* $\mathsf{BW}^{\omega}_{\mathbb{R}}$*. The term* t^*_L *depends only on M (but not on s) and* t^*_A *even is independent from both s and M, indeed:*

$$t^*_L(M) :=_1 \lambda n^0.\Phi^*_0(X^*M)u^{*1(1(0))}_{(Z^*M)}0^0 0^1(\bar{\mathbf{1}}(\mathrm{lv}^*(n))),$$

$$t^*_A :=_{1(0)} \lambda d^0, n^0.N_{K_d},$$

where

$$N_m(k) := j(m2^{k+3}+1, 2^{k+2}-1)$$

and X^*M *and* Z^*M *are defined primitive recursively in M:*

$$X^*M := \lambda g^1 . \bar{\mathbf{1}}(M'g), \qquad Z^*M := \lambda g^1 . \max_0(\, M'g,\ g(\bar{\mathbf{1}}(M'g))\,).$$

The term M' *is a similar primitive recursive modification of M as before:*

$$M'g := \mathrm{lv}^*\Big(M\left(\lambda n^0.g(\bar{\mathbf{1}}(\mathrm{lv}^*(n)))\right)(t^*_A)\Big).$$

The majorized versions of u and lv*, are given by:*

$$u^*_Z n^0 v^{1(0)} := \max_0(1, Z(v1)), \quad \mathrm{lv}^*(n) := (\max_{d\leq n+1}{}_{\mathbb{N}}\{K_d, n\}+2)^4.$$

Though it is not entirely obvious how we obtain this theorem, the actual steps are purely elementary.

Similarly, we can obtain the majorized version of Theorem 4.7:

Theorem 4.11. *The terms* t^*_F *and* t^*_P *satisfy the monotone Sh-interpretation of* $\mathsf{BW}^{\omega}_{\mathbb{R}}$ *stating the existence of a converging subsequence (see also Theorem 4.7 above):*

$$t^*_F M := \lambda n.(t^*_L M'')^n 0, \quad t^*_P := t^*_A,$$

with t^*_L *and* t^*_A *defined as in the Theorem 4.10 and the term and where the term* M'' *is a similar primitive recursive modification of M as in Theorem 4.7:*

$$M''la := l^{M(\lambda n.l^n(0))a}(0).$$

Proof. For M^* s-maj M, we have that $(M^*)''$ s-maj M', M''. Hence $t^*_L((M^*)'')(n) \geq t_L sM'(n)$ and the function $t^*_L((M^*)'')$ is non-decreasing. Therefore $t^*_F M^*$ is non-decreasing as well and majorizes $t_F sM$. □

4.2 Analysis of the Complexity of the Realizers

Theorem 4.10 implies that we can use only a single application of $\mathsf{B}_{0,1}$ on primitive recursive functionals and primitive recursion to obtain the realizing terms for $\mathsf{BW}^{\omega}_{\mathbb{R}}$.
Now, we can investigate how the principle $\mathsf{BW}^{\omega}_{\mathbb{R}}$ does affect the complexity of the realizers of a given theorem proved using this principle. Depending on the way the $\mathsf{BW}^{\omega}_{\mathbb{R}}$-principle is used in such a proof, we get the results stated in theorems 4.12 and 4.15 respectively.

Theorem 4.12. Program extraction for proofs based on an instance of $\mathsf{BW}^{\omega}_{\mathbb{R}}$.
For $n \geq 1$, given a proof using $\mathsf{BW}^{\omega}_{\mathbb{R}}$ on a known sequence sx of elements in $\mathbb{P}$ specified by a closed term s of $\widehat{\mathsf{WE\text{-}PA}}^{\omega}\restriction$, i.e. $s \in (\mathbb{P}^{\mathbb{N}})^{\mathbb{N}}$ defines a sequence of such sequences (where the bounds (k_d) used in forming $\mathbb{P}$ are also given by a term rx that may depend on x), and a quantifier-free formula $\varphi_{\mathsf{QF}}(x,y)$ containing only x,y as free variables we have: from a proof

$$\widehat{\mathsf{WE\text{-}PA}}^{\omega}\restriction + \mathsf{QF\text{-}AC} + \Sigma^0_n\text{-}\mathsf{IA} \vdash \forall x^0(\mathsf{BW}^{\omega}_{\mathbb{R}}(sx) \to \exists y^0 \varphi_{\mathsf{QF}}(x,y)),$$

we can extract a function $f \in T_n$ by Sh-interpretation s.t.

$$\mathcal{S}^{\omega} \models \forall x^0 \varphi_{\mathsf{QF}}(x, f(x)).$$

Remark. 1. In particular, for $n = 1$ we obtain a $<\omega^{\omega^{\omega}}$- recursive realizer. Moreover, for the special case $n = 0$, U: Kohlenbach showed in [64] that when some weaker systems than $\widehat{\mathsf{WE\text{-}PA}}^{\omega}\restriction + \mathsf{QF\text{-}AC}$ are used, namely $\mathsf{G}_{\infty}\mathsf{A}^{\omega} + \mathsf{QF\text{-}AC}$, one obtains even $<\omega^{\omega}$-recursive realizers. Note that this case is not covered by Theorem 4.12 above.

2. Instead of $\widehat{\mathsf{WE\text{-}PA}}^{\omega}\restriction$ we may also have the system resulting from this by adding the recursors R_{ρ} with $deg(\rho) \leq n-1$.

Corollary 4.13. *Theorem 4.12 also holds with $\widehat{\mathsf{E\text{-}PA}}^{\omega}\restriction$ instead of $\widehat{\mathsf{WE\text{-}PA}}^{\omega}\restriction$ once we restrict* QF-AC *to the types $(\rho,\tau) := (1,0)$ and $(\rho,\tau) := (0,1)$.*

Proof. Apply elimination of extensionality to theorem 4.12 (see [93] or [71]). □

Proof of Theorem 4.12. By soundness of $\widehat{\mathsf{WE\text{-}PA}}^{\omega}\restriction$ (see sections 1.1, 1.2 or e.g. [71] and [118]) we know that one can construct closed terms t_M and t_Y in T_{n-1} realizing the Sh-interpretation of the analyzed proof:

$$\forall x, L, a\,(\mathsf{BW}_{Sh}(a, t_M xLa, L(t_M xLa), sx) \to \varphi_{Sh}(x, t_Y xLa)).$$

Now, using the solution terms t_A, t_L for the Shoenfield interpretation of $\mathsf{BW}^{\omega}_{\mathbb{R}}$ from theorem 4.6 we define

$$a := t_A(sx, t_M x) \text{ and } L := t_L(sx, t_M x)$$

and obtain that

$$\mathsf{BW}_{Sh}(a, t_M xLa, L(t_M xLa), sx).$$

Hence for

$$tx := t_Y xLa = t_Y x(t_L(sx, t_M x))(t_A(sx, t_M x))$$

we have that $\varphi_{qf}(x, tx)$.
Now let t^*_M, t^*_Y be majorants for t_M, t_Y in T_{n-1}. Then –using the majorants t^*_L, t^*_A of t_L, t_A from theorem 4.10 – we define

$$t^*x := t^*_Y x(t^*_L(t^*_M x))(t^*_A(t^*_M x)).$$

Since x maj$_0$ x we get that $t^*x \geq tx$ for all x. It, therefore, suffices to show that the function denoted by t^* can be defined in T_n : Using Parsons' result from [103] (p. 361) we know that $t^*_M x$ has computational sizes $< \omega_n(\omega)$. This fact allows us to apply

Howard's proposition, see [53] (p. 23), from which it follows that $t^*_L(t^*_M x)y^0$ has computational size $< \omega_{n+1}(\omega)$ and so (using Parson's result again) can be defined by a term in T_n which finishes the proof as t^*_A even is in T_0.

□

The Theorem 4.12 is optimal in the following sense:

Proposition 4.14. *Any function h given by a closed term in $\mathcal{T}_n$ can be proven to be total in* $\widehat{\text{WE-PA}}^{\omega}\!\upharpoonright + \text{QF-AC} + \Sigma^0_n\text{-IA}$ *using a concrete instance of* $\mathsf{BW}(s)$ *for a suitable closed term* s^1 *in* $\mathcal{T}_0$.

Proof.

In [68] (Proposition 5.5), U. Kohlenbach gave a construction of a functional F^2 such that (relative to $\widehat{\text{WE-PA}}^{\omega}\!\upharpoonright$) $\forall f^{1(0)}(\mathsf{PCM}(F(f)) \to \Pi^0_1\text{-}\widehat{\mathsf{CA}}(f))$, where $\mathsf{PCM}(f^{1(0)})$ is the principle of monotone convergence defined as follows:

$$\forall n(0 \leq_{\mathbb{R}} f(n+1) \leq_{\mathbb{R}} f(n)) \to \exists g \forall k \forall m_1, m_2 \geq gk \left(|fm_1 -_{\mathbb{R}} fm_2| \leq \frac{1}{k+1} \right).$$

Moreover, in the presence of QF-AC we also have $\forall f^{1(0)}(\mathsf{BW}_{\mathbb{R}}(G(f)) \to \mathsf{PCM}(f))$ for a suitable functional G. Hence there are functionals $F_\exists$, and $F_\forall$ such that (relative to $\widehat{\text{WE-PA}}^{\omega}\!\upharpoonright$ + QF-AC) for any function $f^{1(0)}$ the instance $\mathsf{BW}(F_\exists f)$, resp. $\mathsf{BW}(F_\forall f)$, of BW implies the instance $\Sigma^0_1\text{-CA}(f)$ of Σ^0_1-CA, resp. $\Pi^0_1\text{-CA}(f)$ of Π^0_1-CA.
By Parsons's results in [104] we know that $\widehat{\text{WE-PA}}^{\omega}\!\upharpoonright$ + QF-AC proves the Π^0_2 sentence stating the totality of h using additionally just finitely many, say m, instances of Σ^0_{n+1}-IA with number parameters only.
Assume that n is odd. Each of these m instances can be reformulated as an instance of Σ^0_n-IA with number and function parameters plus the instances $\forall \underline{a}^0 \Pi^0_1\text{-CA}(t'_i(\underline{a}))$ of Π^0_1-comprehension, where the term $t'_i \in \mathcal{T}_0$ ($i \in \{1 \dots m\}$) corresponds to the quantifier-free matrix of the induction formula in the ith instance of Σ^0_{n+1}-IA (i.e. t'_i essentially is the characteristic term of that matrix). Furthermore, for suitable closed terms $t_i \in \mathcal{T}_0$ each formula $\forall \underline{a}^0 \Pi^0_1\text{-CA}(t'_i(\underline{a}))$ is equivalent to an instance $\Pi^0_1\text{-CA}(t_i)$ of Π^0_1-CA. Moreover, $\bigwedge_{i=0}^m \Pi^0_1\text{-CA}(t_i)$ is equivalent to $\Pi^0_1\text{-CA}(t)$ for a suitable $t \in \mathcal{T}_0$ (see [65] for this).
Analogously, if n is even we obtain the equivalence to $\Sigma^0_1\text{-CA}(t)$.
Finally, $\Pi^0_1\text{-CA}(t)$ and $\Sigma^0_1\text{-CA}(t)$ are derivable from the instances $\mathsf{BW}(F_\forall t)$ resp. $\mathsf{BW}(F_\exists t)$ of BW.

□

Theorem 4.12 should be contrasted to the case of proofs based on the full (2nd order closure – also w.r.t. the bounding sequence (k_d) from $\mathbb{P}$ – of the) Bolzano-Weierstraß principle BW where the following (equally optimal) result follows from the literature:

Theorem 4.15. *Given a proof using* $\mathsf{BW}^{\omega}_{\mathbb{R}}$ *on any given sequence in* $\mathbb{P}$*: Let* $\varphi_{\mathsf{QF}}(x,y)$ *be a quantifier-free formula containing only the free variables* x, y*. Then from a proof*

$$\widehat{\text{WE-PA}}^{\omega}\!\upharpoonright + \text{QF-AC} + \Sigma^0_\infty\text{-IA} + \mathsf{BW}^{\omega}_{\mathbb{R}} \vdash \forall x^1 \exists y^0 \varphi_{\mathsf{QF}}(x,y)$$

we can extract by Sh-interpretation a closed term $t^1 \in \mathcal{T}$ *s.t.*

$$\mathcal{S}^{\omega} \models \forall x^0 \varphi_{\mathsf{QF}}(x, t(x)).$$

Proof. Over $\widehat{\mathsf{WE\text{-}PA}}^{\omega}\!\restriction + \mathsf{QF\text{-}AC}$ the schema of arithmetical comprehension CA^0_{ar} clearly implies both $\mathsf{BW}^{\omega}_{\mathbb{R}}$ as well as Σ^0_{∞}-IA. The system $\widehat{\mathsf{WE\text{-}PA}}^{\omega}\!\restriction + \mathsf{QF\text{-}AC} + \mathsf{CA}^0_{ar}$, however, has a *Sh*-interpretation by terms in $\mathcal{T}_0 + \mathsf{B}_{0,1}$ (see e.g. [71], Theorem 11.14). Hence we get a term $t \in \mathcal{T}_0 + \mathsf{B}_{0,1}$ satisfying $\forall x^1 \varphi_{\mathsf{QF}}(x, t(x))$. Finally, using Corollary 4.4.1 from [67], we can conclude that t can be rewritten as a functional in $\mathcal{T}$.

□

Functional interpretation for IST

This paper introduces a new realizability and functional interpretation where throughout the interpretation one works with non-empty finite sequences (sets) of witnesses, rather than single witnesses. This, surprisingly, gives rise to functional interpretations which are extremely suitable to deal with the kind of principles used in non-standard analysis such as overspill, underspill and the idealization principle, amongst others. The proposed set of techniques is much more involved that the usual modifed realizabilty and dialectica interpretations, in that one has to deal with two different kinds of quantifiers (internal and external) and a new atomic predicate $\mathrm{st}(x)$, *and their corresponding defining axioms. The benefit of this is that most of the classical principles of non-standard analysis become either interpretable or are even eliminated by the interpretation, leading to very strong conservation results which extends previous results of Moerdijk and Palmgren (1997) and Avigad and Helzner (2002). This is a well-written paper introducing some novel and powerful ideas in the area of functional interpretation, with some impressive applications to non-standard analysis...*

– anonymous referee about [14], Apr 26, 2012.

Motivation

In this section we give functional interpretations for both constructive and classical systems of nonstandard arithmetic. These results are a first step towards the interpretation of the countable saturation principle (which will hopefully one day lead to a functional interpretation of proofs based on Loeb measures). In particular, we show that the nonstandard systems are conservative over ordinary (standard) ones and we show how terms can be extracted from nonstandard proofs.

Let us have a short look at the work of Nelson on conservation results for nonstandard systems, also because it was a major source of inspiration for [14] (large part of this chapter is based on that article). The idea of Nelson was to add a new unary predicate symbol st to ZFC for an object "being standard". Using this predicate, he added three new axioms to ZFC governing its use and formalizing the basic non-standard principles. He calls these axioms Idealization, Standardization and Transfer resulting in the system he calls IST, which actually stands for Internal Set Theory. His main logical result about IST is that it is a conservative extension of ZFC, so any theorem provable in IST (which does not involve the st-predicate, of course) is provable also in ZFC.

He proved the conservativity twice. In the original paper introducing Internal Set Theory [97] (reprinted in Volume 48, Number 4 of the *Bulletin of the American Mathematical Society* in recognition of its status as a classic), and in a later publication [98]. The latter proof is done syntactically by providing a "reduction algorithm" (a rewriting algorithm) for converting proofs performed in IST to ordinary ZFC-proofs. There is a remarkable similarity between this reduction algorithm and the Shoenfield interpretation [112] (see also Definition 1.26). This observation was the starting point for [14].

Let us point out that [14] shows that if one defines a Dialectica-type functional interpretation using the new application, with implication interpreted *à la* Diller–Nahm [23], it will interpret and eliminate principles recognizable from nonstandard analysis. By combining that functional interpretation with negative translation, we were able to define a Shoenfield-type functional interpretation for classical nonstandard systems as well. In this way we also obtained conservation and term extraction results for classical systems. The resulting functional interpretations in [14] (see sections 5.3 and 5.5) have some striking similarities with the bounded functional interpretations introduced by Ferreira and Oliva in [27] and [26] (see also [29]).

5.1 Formalities

In this section, we follow very closely [14] and extend our first section to cover the necessary technicalities to formalize proofs in non-standard analysis.

The system $\mathsf{E\text{-}HA}^{\omega *}$

In this chapter, $\mathsf{E\text{-}HA}^{\omega *}$ will be the extension of the system called $\mathsf{E\text{-}HA}_0^{\omega}$ in [125] and $\mathsf{E\text{-}HA}_{\rightarrow}^{\omega}$ in [127] with types for finite sequences, see also Definition 1.9 in first chapter (though note that we treat things here a little differently – see below). More precisely, the collection of types $\boldsymbol{T}^*$ (similarly to $\boldsymbol{T}$, see Definition 1.1) will be the smallest set closed under the following rules (note that σ^* denotes a finite sequence of elements of type σ):

(i) $0 \in \boldsymbol{T}^*$;

(ii) $\sigma, \tau \in \boldsymbol{T}^* \Rightarrow (\sigma \to \tau) \in \boldsymbol{T}^*$;

(iii) $\sigma \in \boldsymbol{T}^* \Rightarrow \sigma^* \in \boldsymbol{T}^*$.

In dealing with tuples, we will follow the notation and conventions of the first chapter, [125] and [71]. Specifically for this context see also Details in [14]. This includes the enrichment of the term language (Gödel's $\mathcal{T}$); it now also includes a constant $\langle\rangle_\sigma$ of type σ^* and an operation c of type $\sigma \to (\sigma^* \to \sigma^*)$ (for the empty sequence and the operation of prepending an element to a sequence, respectively), as well as a list recursor $\underline{L}_{\sigma,\rho}$ satisfying the usual axioms (again see [14] and also [127, p. 456] or [71, p. 48]). In addition, we have the recursors and combinators for all the new types in Gödel's $\mathcal{T}$, satisfying the usual equations. The resulting extension we will denote by $\mathcal{T}^*$.

Differently to previous chapters, we will have a primitive notion of equality at every type and equality axioms expressing that equality is a congruence (as in [127, p. 448-9]). Since decidability of quantifier-free formulas is not essential for this chapter, this choice will not create any difficulties. In addition, we assume the axiom of extensionality for functions:

$$f =_{\sigma\to\tau} g \leftrightarrow \forall x^\sigma\, fx =_\tau gx.$$

The axiom schemas of the underlying system (i.e. $\mathsf{E\text{-}HA}^\omega$) apply to all formulas in the language (i.e., also those containing variables of sequence type and the new terms that belong to $\mathcal{T}^*$).
Finally, we add the following sequence axiom:

$$\mathsf{SA}: \quad \forall y^{\sigma^*}\,(\, y = \langle\rangle_\sigma \vee \exists a^\sigma, x^{\sigma^*}\; y = c(a,x)\,).$$

In the usual formalization of $\mathsf{E\text{-}HA}^\omega$, as in [71] or [127] (or simply our system from Definition 1.9), for example, one can also talk about sequences, but these have to be coded up (see [71, p. 59]). As a result, $\mathsf{E\text{-}HA}^{\omega*}$ is a definitional extension of, and hence conservative over, $\mathsf{E\text{-}HA}^\omega$ as defined in [71] or [127].

The system $\mathsf{E\text{-}HA}^{\omega*}_{\mathrm{st}}$

Definition 5.1 (As given in [14]). The language of the system $\mathsf{E\text{-}HA}^{\omega*}_{\mathrm{st}}$ is obtained by extending that of $\mathsf{E\text{-}HA}^{\omega*}$ with unary predicates st^σ as well as two new quantifiers $\forall^{\mathrm{st}} x^\sigma$ and $\exists^{\mathrm{st}} x^\sigma$ for every type $\sigma \in \boldsymbol{T}^*$. Formulas in the language of $\mathsf{E\text{-}HA}^{\omega*}$ (i.e., those that do not contain the new predicate st_σ or the two new quantifiers $\forall^{\mathrm{st}} x^\sigma$ and $\exists^{\mathrm{st}} x^\sigma$) will be called *internal, denoted – as before – by small Greek letters e.g. φ, ψ*. Formulas which are not internal will be called *external, denoted by capital Greek letters, e.g. Φ, Ψ*.

Definition 5.2 ($\mathsf{E\text{-}HA}^{\omega*}_{\mathrm{st}}$, in [14]). The system $\mathsf{E\text{-}HA}^{\omega*}_{\mathrm{st}}$ is obtained by adding to $\mathsf{E\text{-}HA}^{\omega*}$ the axioms EQ, $\mathcal{T}^*_{\mathrm{st}}$ and $\mathsf{IA}^{\mathrm{st}}$, where

- EQ stands for the defining axioms of the external quantifiers:

$$\begin{aligned} \forall^{\mathrm{st}} x\, \Phi(x) &\leftrightarrow \forall x\,(\,\mathrm{st}(x) \to \Phi(x)\,), \\ \exists^{\mathrm{st}} x\, \Phi(x) &\leftrightarrow \exists x\,(\,\mathrm{st}(x) \wedge \Phi(x)\,), \end{aligned}$$

  with $\Phi(x)$ an arbitrary formula, possibly with additional free variables.

- $\mathcal{T}^*_{\mathrm{st}}$ consists of:

  1. the axioms $\mathrm{st}(x) \wedge x = y \to \mathrm{st}(y)$,
  2. the axiom $\mathrm{st}(t)$ for each closed term t in $\mathcal{T}^*$,
  3. the axioms $\mathrm{st}(f) \wedge \mathrm{st}(x) \to \mathrm{st}(fx)$.

- $\mathsf{IA}^{\mathrm{st}}$ is the external induction axiom:

$$\mathsf{IA}^{\mathrm{st}} \quad : \quad (\Phi(0) \wedge \forall^{\mathrm{st}} n^0(\Phi(n) \to \Phi(n+1))) \to \forall^{\mathrm{st}} n^0 \Phi(n),$$

  where $\Phi(n)$ is an arbitrary formula, possibly with additional free variables.

Here it is to be understood that in $\mathsf{E\text{-}HA}^{\omega*}_{\mathrm{st}}$ the laws of intuitionistic logic apply to all formulas, while the induction axiom from $\mathsf{E\text{-}HA}^{\omega*}$

$$(\varphi(0) \wedge \forall n^0(\varphi(n) \to \varphi(n+1))) \to \forall n^0 \varphi(n)$$

applies to internal formulas φ only.

Lemma 5.3 ([14]). $\mathsf{E\text{-}HA}^{\omega *}_{\mathrm{st}} \vdash \Phi(x) \wedge x = y \to \Phi(y)$ *for every formula* Φ.

Lemma 5.4 ([14]). $\mathsf{E\text{-}HA}^{\omega *}_{\mathrm{st}} \vdash \mathrm{st}^0(x) \wedge y \leq x \to \mathrm{st}^0(y)$.

Definition 5.5 ([14]). For any formula Φ in the language of $\mathsf{E\text{-}HA}^{\omega *}_{\mathrm{st}}$, we define its *internalization* Φ^{int} to be the formula one obtains from Φ by replacing $\mathrm{st}(x)$ by $x = x$, and $\forall^{\mathrm{st}} x$ and $\exists^{\mathrm{st}} x$ by $\forall x$ and $\exists x$, respectively.

One of the reasons $\mathsf{E\text{-}HA}^{\omega *}_{\mathrm{st}}$ is such a convenient system for the proof-theoretic investigations in [14] is because we have the following easy result:

Proposition 5.6 ([14]). *If a formula* Φ *is provable in* $\mathsf{E\text{-}HA}^{\omega *}_{\mathrm{st}}$*, then its internalization* Φ^{int} *is provable in* $\mathsf{E\text{-}HA}^{\omega *}$. *Hence* $\mathsf{E\text{-}HA}^{\omega *}_{\mathrm{st}}$ *is a conservative extension of* $\mathsf{E\text{-}HA}^{\omega *}$ *and* $\mathsf{E\text{-}HA}^{\omega}$.

Operations on finite sequences

We have all the standard operations on finite sequences, see [14] for details and the following lemma.

Lemma 5.7 ([14]).

1. $\mathsf{E\text{-}HA}^{\omega *}_{\mathrm{st}} \vdash \mathrm{st}(x^{\sigma^*}) \to \mathrm{st}(|x|)$,
2. $\mathsf{E\text{-}HA}^{\omega *}_{\mathrm{st}} \vdash \mathrm{st}(x^{\sigma^*}) \to \mathrm{st}((x)_i)$,
3. $\mathsf{E\text{-}HA}^{\omega *}_{\mathrm{st}} \vdash \mathrm{st}(x_0^{\sigma}) \wedge \ldots \wedge \mathrm{st}(x_n^{\sigma}) \to \mathrm{st}(\langle x_0^{\sigma}, \ldots, x_n^{\sigma} \rangle)$,
4. $\mathsf{E\text{-}HA}^{\omega *}_{\mathrm{st}} \vdash \mathrm{st}(x^{\sigma^*}) \wedge \mathrm{st}(y^{\sigma^*}) \to \mathrm{st}(x *_{\sigma} y)$.
5. $\mathsf{E\text{-}HA}^{\omega *}_{\mathrm{st}} \vdash \mathrm{st}(F^{0 \to \sigma^*}) \wedge \mathrm{st}(n^0) \to \mathrm{st}(F(0) * \ldots * F(n-1))$.

Proof. Follows from the $\mathcal{T}^*_{\mathrm{st}}$-axioms together with the fact that the list recursor L belongs to $\mathcal{T}^*$. □

Finite sets

Most of the time, as in [14], we will regard finite sequences as stand-ins for finite sets. We also use the notion of an element and that of one sequence being contained in another, as given in [14].

Definition 5.8 ([14]). For s^{σ}, t^{σ^*} we write $s \in_{\sigma} t$ and say that s *is an element of* t if

$$\exists i < |t| (s =_{\sigma} (t)_i).$$

For $\underline{s}^{\underline{\sigma}} = s_0^{\sigma_0}, \ldots, s_{n-1}^{\sigma_{n-1}}$ and $\underline{t}^{\underline{\sigma}^*} = t_0^{\sigma_0^*}, \ldots, t_{n-1}^{\sigma_{n-1}^*}$ we write $\underline{s} \in_{\underline{\sigma}} \underline{t}$ and say that $\underline{s}$ *is an element of* $\underline{t}$ if

$$\bigwedge_{k=0}^{n-1} s_k \in_{\sigma_k} t_k.$$

In case no confusion can arise, we will drop the subscript and write simply $\in$ instead of $\in_{\sigma}$ or $\in_{\sigma^*}$.

Lemma 5.9 ([14]). $\mathsf{E\text{-}HA}^{\omega *}_{\mathrm{st}} \vdash \mathrm{st}(x^{\sigma^*}) \wedge y \in_{\sigma} x \to \mathrm{st}(y^{\sigma})$.

Definition 5.10 ([14]). For $s^{\sigma^*}, t^{\sigma^*}$ we write $s \preceq_\sigma t$ and say that s *is contained in* t if

$$\forall x^\sigma\, (\, x \in s \to x \in t\,),$$

or, equivalently,

$$\forall i < |s|\, \exists j < |t|\, (s)_i =_\sigma (t)_j.$$

For $\underline{s}^{\underline{\sigma}^*} = s_0^{\sigma_0^*}, \ldots, s_{n-1}^{\sigma_{n-1}^*}$ and $\underline{t}^{\underline{\sigma}^*} = t_0^{\sigma_0^*}, \ldots, t_{n-1}^{\sigma_{n-1}^*}$ we write $\underline{s} \preceq_{\underline{\sigma}} \underline{t}$ and say that $\underline{s}$ *is contained in* $\underline{t}$ if

$$\bigwedge_{k=0}^{n-1} s_k \preceq_{\sigma_k} t_k.$$

Lemma 5.11 ([14]). E-HA$^{\omega*}$ *proves that* $\preceq_\sigma$ *determines a preorder on the set of objects of type* σ^**. More precisely, for all* x^{σ^*} *we have* $x \preceq_\sigma x$*, and for all* $x^{\sigma^*}, y^{\sigma^*}, z^{\sigma^*}$ *with* $x \preceq_\sigma y$ *and* $y \preceq_\sigma z$*, we have* $x \preceq_\sigma z$.

Definition 5.12 ([14]). A property $\Phi(\underline{x}^{\underline{\sigma}^*})$ is called *upwards closed in* $\underline{x}$ if $\Phi(\underline{x}) \wedge \underline{x} \preceq \underline{y} \to \Phi(\underline{y})$ and *downwards closed in* $\underline{x}$ if $\Phi(\underline{x}) \wedge \underline{y} \preceq \underline{x} \to \Phi(\underline{y})$.

Induction and extensionality for sequences

Proposition 5.13 ([14]). E-HA$^{\omega*}$ *proves the induction schema for sequences:*

$$\varphi(\langle\rangle_\sigma) \wedge \forall a^\sigma, y^{\sigma^*}\, (\, \varphi(y) \to \varphi(c(a,y))\,) \to \forall x^{\sigma^*}\, \varphi(x).$$

A consequence of this is the principle of extensionality for sequences. We follow [14] and call two elements $x^{\sigma^*}, y^{\sigma^*}$ extensionally equal, and write $x =_{e,\sigma^*} y$, iff

$$|x| =_0 |y| \wedge \forall i < |x|\, (\, (x)_i =_\sigma (y)_i\,).$$

Proposition 5.14 ([14]). E-HA$^{\omega*}$ *proves*

$$\forall x^{\sigma^*}, y^{\sigma^*}\, (\, x =_{e,\sigma^*} y \to x =_{\sigma^*} y\,).$$

Corollary 5.15 ([14]). E-HA$^{\omega*}_{\mathrm{st}}$ *proves*

$$\forall x^{\sigma^*}\, \mathrm{st}(|x|) \wedge \forall i < |x|\, \mathrm{st}((x)_i) \to \mathrm{st}(x).$$

Corollary 5.16 ([14]). E-HA$^{\omega*}_{\mathrm{st}}$ *proves the external induction axiom for sequences:*

$$\Phi(\langle\rangle_\sigma) \wedge \forall^{\mathrm{st}} a^\sigma, y^{\sigma^*}\, (\, \Phi(y) \to \Phi(c(a,y))\,) \to \forall^{\mathrm{st}} x^{\sigma^*}\, \Phi(x).$$

Finite sequence application

We already said, that we have all the usual operations on sequences. The following, more involved, operations are crucial for this chapter.

Definition 5.17 (Finite sequence application and abstraction, [14]). If s is of type $(\sigma \to \tau^*)^*$ and t is of type σ, then

$$s[t] := (s)_0(t) * \ldots * (s)_{|s|-1}(t) : \tau^*.$$

For every term s of type $\sigma \to \tau^*$ we set

$$\Lambda x^\sigma.s(x) := \langle \lambda x^\sigma.s(x)\rangle : (\sigma \to \tau^*)^*.$$

Note that we have

$$(\Lambda x.s(x))[t] =_{\tau^*} (\lambda x.s(x))(t) =_{\tau^*} s(t).$$

Also, the same conventions as for ordinary application and abstraction apply.

Moreover, in [14] we also define recursors $\underline{\mathcal{R}_\rho}$ for each tuple of types $\underline{\rho}^* = \rho_0^*, \ldots, \rho_k^*$, such that

$$\begin{aligned} \underline{\mathcal{R}_\rho}(0, \underline{y}, \underline{z}) &=_{\underline{\rho}^*} \underline{y}, \\ \underline{\mathcal{R}_\rho}(n+1, \underline{y}, \underline{z}) &=_{\underline{\rho}^*} \underline{z}[n, \underline{\mathcal{R}_\rho}(n, \underline{y}, \underline{z})], \end{aligned}$$

(where y_i is of type ρ_i^* and z_i is of type $(0 \to \rho_0^* \to \ldots \to \rho_k^* \to \rho_i^*)^*$). Indeed, by letting

$$\underline{\mathcal{R}_\rho} := \lambda n^0, \underline{y}, \underline{z}.\underline{R}_{\rho^*}(n, \underline{y}, (\lambda \underline{s}^{\underline{\rho}^*}, t^0.\underline{z}[t, \underline{s}])),$$

where $\underline{R_\rho}$ are constants for simultaneous primitive recursion as in [71], we get

$$\underline{\mathcal{R}_\rho}(0, \underline{y}, \underline{z}) =_{\underline{\rho}^*} \underline{R}_{\rho^*}(0, \underline{y}, (\lambda \underline{s}^{\underline{\rho}^*}, t^0.\underline{z}[t, \underline{s}])) =_{\underline{\rho}^*} \underline{y}$$

and

$$\begin{aligned} \underline{\mathcal{R}_\rho}(n+1, \underline{y}, \underline{z}) &=_{\underline{\rho}^*} \underline{R}_{\rho^*}(n+1, \underline{y}, (\lambda \underline{s}^{\underline{\rho}^*}, t^0.\underline{z}[t, \underline{s}])) \\ &=_{\underline{\rho}^*} (\lambda \underline{s}^{\underline{\rho}^*}, t^0.\underline{z}[t, \underline{s}])(\underline{R}_{\rho^*}(n, \underline{y}, (\lambda \underline{s}^{\underline{\rho}^*}, t^0.\underline{z}[t, \underline{s}])), n) \\ &=_{\underline{\rho}^*} \underline{z}[n, \underline{R}_{\rho^*}(n, \underline{y}, (\lambda \underline{s}^{\underline{\rho}^*}, t^0.\underline{z}[t, \underline{s}]))] \\ &=_{\underline{\rho}^*} \underline{z}[n, \underline{\mathcal{R}_\rho}(n, \underline{y}, \underline{z})]. \end{aligned}$$

We have the following concerning the preorder defined above.

Lemma 5.18 ([14]). E-HA$^{\omega*}$ *proves*

1. *If* $s^{(\sigma\to\tau^*)^*} \preceq \tilde{s}^{(\sigma\to\tau^*)^*}$, *then* $s[t] \preceq \tilde{s}[t]$, *for all* t^σ.
2. *If* $s \preceq \tilde{s}$, *then* $s[\underline{t}] \preceq \tilde{s}[\underline{t}]$ *for all* $\underline{t}$ *of suitable types.*
3. *If* $\underline{s} \preceq \underline{\tilde{s}}$, *then* $\underline{s}[\underline{t}] \preceq \underline{\tilde{s}}[\underline{t}]$ *for all* $\underline{t}$ *of suitable types.*

Lemma 5.19 ([14]). E-HA$^{\omega*}_{\text{st}}$ *proves*

$$\text{st}^{(\sigma\to\tau^*)^*}(x) \wedge \text{st}^\sigma(y) \to \text{st}^{\tau^*}(x[y])$$

and

$$\text{st}^{\sigma\to\tau^*}(s) \to \text{st}^{(\sigma\to\tau^*)^*}(\Lambda x^\sigma.s(x)).$$

5.2 Nonstandard principles

Let us motivate this section with a similar introduction as in [14].
Nonstandard analysis employs the existence of nonstandard models of the first-order theory of the reals (or the natural numbers). One uses the compactness theorem for first-order logic (or, alternatively, the existence of suitable nonprincipal ultrafilters) to show that there are extensions of the natural numbers (or the reals, or other structures)

that are *elementary*, i.e. satisfy the same first-order sentences (or formulas with parameters from the original structure). E.g. for the natural numbers, this means that there are structures $^*\mathbb{N}$ and embeddings $i : \mathbb{N} \to {}^*\mathbb{N}$ that satisfy

$$^*\mathbb{N} \vDash \varphi(i(n_0), \ldots, i(n_k)) \iff \mathbb{N} \vDash \varphi(n_0, \ldots, n_k)$$

for all first-order formulas $\varphi(x_0, \ldots, x_k)$ and natural numbers $n_0, \ldots, n_k$.
The image of i is then called the *standard* natural numbers, while those that do not lie in the image of i are the *nonstandard* natural numbers. Also, it is common to add a new predicate st to the structure $^*\mathbb{N}$, which is true only for the standard natural numbers.
The elementarity of the embedding implies that $^*\mathbb{N}$ is still a linear order in which the nonstandard natural numbers must be infinite (i.e., bigger than any standard natural number). The point of nonstandard systems is that one can use these infinite natural numbers to prove theorems in the nonstandard structure $^*\mathbb{N}$, which must then be true in $\mathbb{N}$ as well, since the embedding i is elementary. The same applies to the reals, with the addition of infinitesimals (nonstandard reals having an absolute value smaller than any positive standard real). Typically, these infinitesimals are then used to prove theorems in analysis in $^*\mathbb{R}$ to show that they must hold in $\mathbb{R}$ as well.
Of course, this makes sense only *first-order, internal* statements. So, using nonstandard models requires some understanding about what can and what can not be expressed in first-order logic as well as whether formulas are internal.

Let us the following, most important principles in nonstandard analysis:

1. Overspill: if $\varphi(x)$ is internal and holds for all standard x, then $\varphi(x)$ also holds for some nonstandard x.

2. Underspill: if $\varphi(x)$ is internal and holds for all nonstandard x, then $\varphi(x)$ also holds for some standard x.

3. Transfer: an internal formula φ (possibly with standard parameters) holds in $^*\mathbb{N}$ iff it holds in $\mathbb{N}$.

These principles will provide us with three criteria with which we will be able to measure the success of the different interpretations. Let us have a closer look.

Remark. Unless we state otherwise, the formulas in this chapter may have additional parameters besides those explicitly shown.

Overspill

When formalised in $\mathsf{E\text{-}HA}^{\omega*}_{\mathrm{st}}$, overspill (in type 0) is the following statement:

$$\mathsf{OS}_0 : \forall^{\mathrm{st}} x^0\, \varphi(x) \to \exists x^0\, (\neg\, \mathrm{st}(x) \wedge \varphi(x)\,).$$

Proposition 5.20. [102] *In* $\mathsf{E\text{-}HA}^{\omega*}_{\mathrm{st}}$*, the principle* OS_0 *implies the existence of nonstandard natural numbers,*

$$\mathsf{ENS}_0 : \exists x^0\, \neg\, \mathrm{st}(x),$$

as well as:

$$\mathsf{LLPO}^{\mathrm{st}}_0 : \forall^{\mathrm{st}} x^0, y^0\, (\,\varphi(x) \vee \psi(y)\,) \to \forall^{\mathrm{st}} x^0\, \varphi(x) \vee \forall^{\mathrm{st}} y^0\, \psi(y).$$

Formulated for all types, we get:

$$\mathsf{OS} : \forall^{\mathrm{st}} x^\sigma\, \varphi(x) \to \exists x^\sigma\, (\neg\, \mathrm{st}(x) \wedge \varphi(x)\,).$$

Generalized, this becomes a higher-type version of Nelson's idealization principle [97]:

$$\mathsf{I} : \forall^{\mathrm{st}} x^{\sigma^*} \exists y^{\tau} \forall x' \in_{\sigma} x \, \varphi(x', y) \to \exists y^{\tau} \forall^{\mathrm{st}} x^{\sigma} \, \varphi(x, y).$$

Proposition 5.21. [102] *In* $\mathsf{E\text{-}HA}^{\omega *}_{\mathrm{st}}$, *the idealization principle* I *implies overspill, as well as the statement that for every type* σ *there is a nonstandard sequence containing all the standard elements of that type:*

$$\mathsf{USEQ} : \exists y^{\sigma^*} \forall^{\mathrm{st}} x^{\sigma} \, x \in_{\sigma} y.$$

Proposition 5.22 ([14]). *In* $\mathsf{E\text{-}HA}^{\omega *}_{\mathrm{st}}$, *the idealization principle* I *implies the existence of nonstandard elements of any type,*

$$\mathsf{ENS} : \exists x^{\sigma} \neg \, \mathrm{st}(x),$$

as well as $\mathsf{LLPO}^{\mathrm{st}}$ *for any type:*

$$\mathsf{LLPO}^{\mathrm{st}} : \forall^{\mathrm{st}} x^{\sigma}, y^{\sigma} \, (\, \varphi(x) \vee \psi(y) \,) \to \forall^{\mathrm{st}} x^{\sigma} \, \varphi(x) \vee \forall^{\mathrm{st}} y^{\sigma} \, \psi(y).$$

Of course, classically (intuitionistically, things are not so clear), idealization is equivalent to its dual

$$\mathsf{R} : \forall y^{\tau} \exists^{\mathrm{st}} x^{\sigma} \, \varphi(x, y) \to \exists^{\mathrm{st}} x^{\sigma^*} \forall y^{\tau} \exists x' \in x \, \varphi(x', y),$$

which was dubbed the *realization principle* in [14].

Our interpretation for constructive nonstandard analysis actually eliminates the stronger *nonclassical realization principle*:

$$\mathsf{NCR} : \forall y^{\tau} \exists^{\mathrm{st}} x^{\sigma} \, \Phi(x, y) \to \exists^{\mathrm{st}} x^{\sigma^*} \forall y^{\tau} \exists x' \in x \, \Phi(x', y),$$

where $\Phi(x, y)$ can be any formula. This is quite remarkable, as NCR is incompatible with classical logic (hence the name) in that one can prove:

Proposition 5.23 ([14]). *In* $\mathsf{E\text{-}HA}^{\omega *}_{\mathrm{st}}$, *the nonclassical realization principle* NCR *implies the undecidability of the standardness predicate:*

$$\neg \forall x^{\sigma} \, (\, \mathrm{st}(x) \vee \neg \, \mathrm{st}(x) \,).$$

Underspill

Underspill (in type 0) is the following statement:

$$\mathsf{US}_0 : \forall x^0 \, (\, \neg \, \mathrm{st}(x) \to \varphi(x) \,) \to \exists^{\mathrm{st}} x^0 \, \varphi(x).$$

In a constructive context it has the following nontrivial consequence (compare [9]):

Proposition 5.24 ([14]). *In* $\mathsf{E\text{-}HA}^{\omega *}_{\mathrm{st}}$, *the underspill principle* US_0 *implies*

$$\mathsf{MP}^{\mathrm{st}}_0 : (\, \forall^{\mathrm{st}} x^0 \, (\, \varphi(x) \vee \neg \varphi(x) \,) \wedge \neg\neg \exists^{\mathrm{st}} x^0 \varphi(x) \,) \to \exists^{\mathrm{st}} x^0 \varphi(x).$$

In particular, $\mathsf{E\text{-}HA}^{\omega *}_{\mathrm{st}} + \mathsf{US}_0 \vdash \neg\neg \, \mathrm{st}^0(x) \to \mathrm{st}^0(x)$.

Also underspill has a direct generalization to higher types:

$$\mathsf{US} : \forall x^{\sigma} \, (\, \neg \, \mathrm{st}(x) \to \varphi(x) \,) \to \exists^{\mathrm{st}} x^{\sigma} \, \varphi(x).$$

Transfer

Following Nelson [97], the transfer principle is usually formulated as follows:

$$\mathsf{TP}_\forall : \forall^{\mathrm{st}} \underline{t} \, (\forall^{\mathrm{st}} x \, \varphi(x, \underline{t}) \to \forall x \, \varphi(x, \underline{t})),$$

where, this time, x and $\underline{t}$ include all free variables of the formula φ. This is classically, but not intuitionistically, equivalent to the following:

$$\mathsf{TP}_\exists : \forall^{\mathrm{st}} \underline{t} \, (\exists x \, \varphi(x, \underline{t}) \to \exists^{\mathrm{st}} x \, \varphi(x, \underline{t})),$$

where, once again, we do not allow parameters.

Interpreting transfer is very difficult, especially in a constructive context (in fact, Avigad and Helzner have devoted an entire paper [9] to this issue). In [14], we discuss three problems as follows:

1. Transfer principles together with overspill imply instances of the law of excluded middle, as was first shown by Moerdijk and Palmgren in [96]. In our setting we have:

   Proposition 5.25 ([14]). *(a) In* $\mathsf{E\text{-}HA}^{\omega*}_{\mathrm{st}}$*, the combination of* ENS_0 *and* $\mathsf{TP}_\forall$ *implies the law of excluded middle for all internal arithmetical formulas.*

   (b) In $\mathsf{E\text{-}HA}^{\omega*}_{\mathrm{st}}$*, the combination of* USEQ *and* $\mathsf{TP}_\forall$ *implies the law of excluded middle for all internal formulas.*

2. As Avigad and Helzner observe in [9], also the combination of transfer principles with underspill results in a system which is no longer conservative over Heyting arithmetic. More precisely, adding US_0 and $\mathsf{TP}_\forall$, or US_0 and $\mathsf{TP}_\exists$, to $\mathsf{E\text{-}HA}^{\omega*}_{\mathrm{st}}$ results in a system which is no longer conservative over Heyting arithmetic HA. The reason is that there are quantifier-free formulas $A(x)$ such that

   $$\mathsf{HA} \nvdash \neg\neg\exists x \, A(x) \to \exists x \, A(x).$$

   Since one can prove a version of Markov's Principle in $\mathsf{E\text{-}HA}^{\omega*}_{\mathrm{st}} + \mathsf{US}_0$, adding either $\mathsf{TP}_\forall$ or $\mathsf{TP}_\exists$ to it would result in a nonconservative extension of HA (and hence of $\mathsf{E\text{-}HA}^{\omega*}$). We refer to [9] for more details.

3. The last point applies to functional interpretations only. As is well-known, in the context of functional interpretations the axiom of extensionality always presents a serious problem and when developing a functional interpretation of nonstandard arithmetic, the situation is no different. Now, $\mathsf{E\text{-}HA}^{\omega*}_{\mathrm{st}}$ includes an internal axiom of extensionality (as it is part of $\mathsf{E\text{-}HA}^{\omega*}$), but for the functional interpretation that we will introduce in Section 5 that will be harmless. What will be very problematic for us, however, is the following version of the axiom of extensionality: if for two elements f, g of type $\sigma_1 \to (\sigma_2 \to \ldots \to 0))$, we define

   $$f =^{\mathrm{st}} g :\equiv \forall^{\mathrm{st}} x_1^{\sigma_1}, x_2^{\sigma_2}, \ldots \, (f \underline{x} =_0 g \underline{x}),$$

   then extensionality formulated as

   $$\forall^{\mathrm{st}} f \, \forall^{\mathrm{st}} x, y \, (x =^{\mathrm{st}} y \to f x =^{\mathrm{st}} f y)$$

   will have no witness definable in ZFC. But that means that also $\mathsf{TP}_\forall$ can have no witness definable in ZFC: for in the presence of $\mathsf{TP}_\forall$ both versions of extensionality are equivalent.

To cope with this, we follow the route taken in most sources (beginning with [95]), to have transfer not as a principle, but as a *rule*. As we will see, this turns out to be feasible. In fact, we will have two transfer rules (which are not equivalent, not even classically):

$$\frac{\forall^{\mathrm{st}} x\, \varphi(x)}{\forall x\, \varphi(x)}\ \mathsf{TR}_\forall \qquad \frac{\exists x\, \varphi(x)}{\exists^{\mathrm{st}} x\, \varphi(x)}\ \mathsf{TR}_\exists$$

(this time, no special requirements on the parameters).

5.3 A functional interpretation for E-HA$^{\omega*}_{\mathrm{st}}$

In this section we will define and study a functional interpretation for E-HA$^{\omega*}_{\mathrm{st}}$ introduced in [14].

The interpretation

The basic idea of the D_{st}-interpretation (the nonstandard Dialectica interpretation) is to associate to every formula $\Phi(\underline{a})$ a new formula $\Phi(\underline{a})^{D_{\mathrm{st}}} \equiv \exists^{\mathrm{st}}\underline{x}\forall^{\mathrm{st}}\underline{y}\, \varphi_{D_{\mathrm{st}}}(\underline{x}, \underline{y}, \underline{a})$ such that

1. all variables in $\underline{x}$ are of sequence type and
2. $\varphi_{D_{\mathrm{st}}}(\underline{x}, \underline{y}, \underline{a})$ is upwards closed in $\underline{x}$.

We will interpret the standardness predicate st^σ similarly to the case for Herbrand realizability: For a realizer for the interpretation of $\mathrm{st}^\sigma(x)$ we will require a standard finite list $\langle y_0, \ldots, y_n\rangle$ of candidates, one of which must be equal to x.

Definition 5.26 (The D_{st}-interpretation for E-HA$^{\omega*}_{\mathrm{st}}$, [14]). We associate to every formula $\Phi(\underline{a})$ in the language of E-HA$^{\omega*}_{\mathrm{st}}$ (with free variables among $\underline{a}$) a formula $\Phi(\underline{a})^{D_{\mathrm{st}}} \equiv \exists^{\mathrm{st}}\underline{x}\forall^{\mathrm{st}}\underline{y}\, \varphi_{D_{\mathrm{st}}}(\underline{x}, \underline{y}, \underline{a})$ in the same language (with the same free variables) by:

(i) $\varphi(\underline{a})^{D_{\mathrm{st}}} :\equiv \varphi_{D_{\mathrm{st}}}(\underline{a}) :\equiv \varphi(\underline{a})$ for internal atomic formulas $\varphi(\underline{a})$,

(ii) $\mathrm{st}^\sigma(u^\sigma)^{D_{\mathrm{st}}} :\equiv \exists^{\mathrm{st}} x^{\sigma^*}\, u \in_\sigma x$.

Let $\Phi(\underline{a})^{D_{\mathrm{st}}} \equiv \exists^{\mathrm{st}}\underline{x}\forall^{\mathrm{st}}\underline{y}\, \varphi_{D_{\mathrm{st}}}(\underline{x}, \underline{y}, \underline{a})$ and $\Psi(\underline{b})^{D_{\mathrm{st}}} \equiv \exists^{\mathrm{st}}\underline{u}\forall^{\mathrm{st}}\underline{v}\, \psi_{D_{\mathrm{st}}}(\underline{u}, \underline{v}, \underline{b})$. Then

(iii) $(\Phi(\underline{a}) \wedge \Psi(\underline{b}))^{D_{\mathrm{st}}} :\equiv \exists^{\mathrm{st}}\underline{x}, \underline{u}\forall^{\mathrm{st}}\underline{y}, \underline{v}\, (\varphi_{D_{\mathrm{st}}}(\underline{x}, \underline{y}, \underline{a}) \wedge \psi_{D_{\mathrm{st}}}(\underline{u}, \underline{v}, \underline{b}))$,

(iv) $(\Phi(\underline{a}) \vee \Psi(\underline{b}))^{D_{\mathrm{st}}} :\equiv \exists^{\mathrm{st}}\underline{x}, \underline{u}\forall^{\mathrm{st}}\underline{y}, \underline{v}\, (\varphi_{D_{\mathrm{st}}}(\underline{x}, \underline{y}, \underline{a}) \vee \psi_{D_{\mathrm{st}}}(\underline{u}, \underline{v}, \underline{b}))$,

(v) $(\Phi(\underline{a}) \to \Psi(\underline{b}))^{D_{\mathrm{st}}} :\equiv \exists^{\mathrm{st}}\underline{U}, \underline{Y}\forall^{\mathrm{st}}\underline{x}, \underline{v}\, (\forall \underline{y} \in \underline{Y}[\underline{x}, \underline{v}]\, \varphi_{D_{\mathrm{st}}}(\underline{x}, \underline{y}, \underline{a}) \to \psi_{D_{\mathrm{st}}}(\underline{U}[\underline{x}], \underline{v}, \underline{b}))$.

Let $\Phi(z, \underline{a})^{D_{\mathrm{st}}} \equiv \exists^{\mathrm{st}}\underline{x}\forall^{\mathrm{st}}\underline{y}\, \varphi_{D_{\mathrm{st}}}(\underline{x}, \underline{y}, z, \underline{a})$, with the free variable z not occuring among the $\underline{a}$. Then

(vi) $(\forall z \Phi(z, \underline{a}))^{D_{\mathrm{st}}} :\equiv \exists^{\mathrm{st}}\underline{x}\forall^{\mathrm{st}}\underline{y}\forall z\, \varphi_{D_{\mathrm{st}}}(\underline{x}, \underline{y}, z, \underline{a})$,

(vii) $(\exists z \Phi(z, \underline{a}))^{D_{\mathrm{st}}} :\equiv \exists^{\mathrm{st}}\underline{x}\forall^{\mathrm{st}}\underline{y}\exists z \forall \underline{y}' \in \underline{y}\, \varphi_{D_{\mathrm{st}}}(\underline{x}, \underline{y}', z, \underline{a})$,

(viii) $(\forall^{\mathrm{st}} z \Phi(z, \underline{a}))^{D_{\mathrm{st}}} :\equiv \exists^{\mathrm{st}}\underline{X}\forall^{\mathrm{st}} z, \underline{y}\, \varphi_{D_{\mathrm{st}}}(\underline{X}[z], \underline{y}, z, \underline{a})$,

(ix) $(\exists^{\mathrm{st}} z \Phi(z, \underline{a}))^{D_{\mathrm{st}}} :\equiv \exists^{\mathrm{st}}\underline{x}, z\, \forall^{\mathrm{st}}\underline{y}\, \exists z' \in z\, \forall \underline{y}' \in \underline{y}\, \varphi_{D_{\mathrm{st}}}(\underline{x}, \underline{y}', z', \underline{a})$.

Definition 5.27 ([14]). We say that a formula Φ is a $\forall^{\text{st}}$-formula if $\Phi \equiv \forall^{\text{st}}\underline{x}\,\varphi(\underline{x})$, with $\varphi(\underline{x})$ internal.

Lemma 5.28 ([14]). *Let Φ be a $\forall^{\text{st}}$-formula. Then $\Phi^{D_{\text{st}}} \equiv \Phi$.*

Note that the clause for $\exists^{\text{st}}z$ causes the interpretation to be not idempotent and that realizers are upwards closed:

Lemma 5.29 ([14]). *Let $\Phi(\underline{a})$ be a formula in the language of E-HA$^{\omega*}_{\text{st}}$ with interpretation $\exists^{\text{st}}\underline{x}\forall^{\text{st}}\underline{y}\,\varphi_{D_{\text{st}}}(\underline{x},\underline{y},\underline{a})$. Then the formula $\varphi_{D_{\text{st}}}(\underline{x},\underline{y},\underline{a})$ is provably upwards closed in $\underline{x}$, i.e.,*

$$\text{E-HA}^{\omega*} \vdash \varphi_{D_{\text{st}}}(\underline{x},\underline{y},\underline{a}) \wedge \underline{x} \preceq \underline{x}' \rightarrow \varphi_{D_{\text{st}}}(\underline{x}',\underline{y},\underline{a}).$$

The D_{st}-interpretation will allow us to interpret the nonclassical realization principle NCR, and also both I and HAC. Additionally we will be able to interpret a herbrandized independence of premise principle for formulas of the form $\forall^{\text{st}}x\,\varphi(x)$, and also a herbrandized form of a generalized Markov's principle:

1. HIP$_{\forall^{\text{st}}}$:

$$(\forall^{\text{st}}x\,\varphi(x) \rightarrow \exists^{\text{st}}y\Psi(y)) \rightarrow \exists^{\text{st}}y\,(\forall^{\text{st}}x\,\varphi(x) \rightarrow \exists y' \in y\,\Psi(y')),$$

where $\Psi(y)$ is a formula in the language of E-HA$^{\omega*}_{\text{st}}$ and $\varphi(x)$ is an internal formula. If $\Psi(y)$ is upwards closed in y, then this is equivalent to

$$(\forall^{\text{st}}x\,\varphi(x) \rightarrow \exists^{\text{st}}y\Psi(y)) \rightarrow \exists^{\text{st}}y\,(\forall^{\text{st}}x\,\varphi(x) \rightarrow \Psi(y)).$$

2. HGMPst:

$$(\forall^{\text{st}}x\,\varphi(x) \rightarrow \psi) \rightarrow \exists^{\text{st}}x\,(\forall x' \in x\,\varphi(x') \rightarrow \psi),$$

where $\varphi(x)$ and ψ are internal formulas in the language of E-HA$^{\omega*}_{\text{st}}$. If $\varphi(x)$ is downwards closed in x, then this is equivalent to

$$(\forall^{\text{st}}x\,\varphi(x) \rightarrow \psi) \rightarrow \exists^{\text{st}}x(\varphi(x) \rightarrow \psi).$$

The latter gives us a form of Markov's principle by taking $\psi \equiv 0 =_0 1$ and $\varphi(x) \equiv \neg\varphi_0(x)$ (with $\varphi_0(x)$ internal and quantifier-free), whence the name.

Theorem 5.30 (Soundness of the D_{st}-interpretation, [14]). *Let $\Phi(\underline{a})$ be a formula of E-HA$^{\omega*}_{\text{st}}$ and let Δ_{int} be a set of internal sentences. If*

$$\text{E-HA}^{\omega*}_{\text{st}} + \text{I} + \text{NCR} + \text{HAC} + \text{HGMP}^{\text{st}} + \text{HIP}_{\forall^{\text{st}}} + \Delta_{\text{int}} \vdash \Phi(\underline{a})$$

and $\Phi(\underline{a})^{D_{\text{st}}} \equiv \exists^{\text{st}}\underline{x}\forall^{\text{st}}\underline{y}\,\varphi_{D_{\text{st}}}(\underline{x},\underline{y},\underline{a})$, then from the proof we can extract closed terms $\underline{t}$ in $\mathcal{T}^*$ such that

$$\text{E-HA}^{\omega*} + \Delta_{\text{int}} \vdash \forall\underline{y}\,\varphi_{D_{\text{st}}}(\underline{t},\underline{y},\underline{a}).$$

Remark 5.31. We could define a system E-HA$^{\omega*}_{\text{nst}}$ by adding primitive predicates nst^{σ} ("nonstandard") to E-HA$^{\omega*}_{\text{st}}$ for each finite type σ, along with axioms

$$\forall x^{\sigma}\big(\text{nst}(x) \leftrightarrow \neg\,\text{st}(x)\big).$$

If we then extend the D_{st}-interpretation by

$$\big(\text{nst}^{\sigma}(x^{\sigma})\big)^{D_{\text{st}}} :\equiv \forall^{\text{st}}y^{\sigma}y \neq_{\sigma} x),$$

we get an analogue of Theorem 5.30, since $\big(\mathrm{nst}(x) \to \neg\,\mathrm{st}(x)\big)^{D_{\mathrm{st}}}$ is provably equivalent to

$$\exists^{\mathrm{st}} Y \forall^{\mathrm{st}} z\, (\forall y \in Y[z](y \neq x) \to x \notin z)$$

and $\big(\neg\,\mathrm{st}(x) \to \mathrm{nst}(x)\big)^{D_{\mathrm{st}}}$ to

$$\exists^{\mathrm{st}} Z \forall^{\mathrm{st}} y\, (\forall z' \in Z[y] x \notin z' \to y \neq x)\,,$$

so that we can take $Y[z] := z$ and $Z[y] := \langle\langle y \rangle\rangle$ respectively .

The characteristic principles of the nonstandard functional interpretation

In [14] we proved that the characteristic principles of the nonstandard functional interpretation are I, NCR, HAC, $\mathsf{HIP}_{\forall^{\mathrm{st}}}$, and $\mathsf{HGMP}^{\mathrm{st}}$. For notational simplicity we will let

$$\mathsf{H} := \mathsf{E\text{-}HA}^{\omega *}_{\mathrm{st}} + \mathrm{I} + \mathsf{NCR} + \mathsf{HAC} + \mathsf{HIP}_{\forall^{\mathrm{st}}} + \mathsf{HGMP}^{\mathrm{st}}.$$

Theorem 5.32 (Characterization theorem for the nonstandard functional interpretation, [14]).

1. *For any formula Φ in the language of $\mathsf{E\text{-}HA}^{\omega *}_{\mathrm{st}}$ we have*

$$\mathsf{H} \vdash \Phi \leftrightarrow \Phi^{D_{\mathrm{st}}}.$$

2. *For any formula Ψ in the language of $\mathsf{E\text{-}HA}^{\omega *}_{\mathrm{st}}$ we have: If for all Φ in $\mathcal{L}(\mathsf{E\text{-}HA}^{\omega *}_{\mathrm{st}})$ (with $\Phi^{D_{\mathrm{st}}} \equiv \exists^{\mathrm{st}} \underline{x} \forall^{\mathrm{st}} \underline{y}\, \varphi_{D_{\mathrm{st}}}(\underline{x}, \underline{y})$) the implication*

$$\mathsf{H} + \Psi \vdash \Phi \quad \Longrightarrow \quad \text{there are closed terms } \underline{t} \in \mathcal{T}^* \text{ s.t. } \mathsf{E\text{-}HA}^{\omega} \vdash \forall \underline{y}\, \varphi_{D_{\mathrm{st}}}(\underline{t}, \underline{y}) \qquad (9)$$

holds, then $\mathsf{H} \vdash \Psi$.

Theorem 5.30 allows us to extract a finite sequence of candidates for the existential quantifier in formulas of the form $\forall^{\mathrm{st}} x\, \exists^{\mathrm{st}} y\, \varphi(x, y)$, in the following sense:

Theorem 5.33 (Main theorem on program extraction by the D_{st}-interpretation, [14]). *Let $\forall^{\mathrm{st}} x \exists^{\mathrm{st}} y\, \varphi(x, y)$ be a sentence of $\mathsf{E\text{-}HA}^{\omega *}_{\mathrm{st}}$ with $\varphi(x, y)$ an internal formula, and let Δ_{int} be a set of internal sentences. If*

$$\mathsf{E\text{-}HA}^{\omega *}_{\mathrm{st}} + \mathrm{I} + \mathsf{NCR} + \mathsf{HAC} + \mathsf{HGMP}^{\mathrm{st}} + \mathsf{HIP}_{\forall^{\mathrm{st}}} + \Delta_{\mathrm{int}} \vdash \forall^{\mathrm{st}} x\, \exists^{\mathrm{st}} y\, \varphi(x, y),$$

then from the proof we can extract a closed term t in $\mathcal{T}^*$ such that

$$\mathsf{E\text{-}HA}^{\omega *} + \Delta_{\mathrm{int}} \vdash \forall x\, \exists y \in t(x)\, \varphi(x, y).$$

It follows from the soundness of the D_{st}-interpretation (Theorem 5.30) that it can be used to eliminate nonstandard principles, like overspill, realization and idealization, from proofs. It also allows one to eliminate underspill, since we have the following result (recall that R is the realization principle from Section 4.1):

Proposition 5.34 ([14]). *We have*

$$\mathsf{E\text{-}HA}^{\omega *}_{\mathrm{st}} + \mathsf{R} + \mathsf{HGMP}^{\mathrm{st}} \vdash \mathsf{US},$$

and therefore the underspill principle US is eliminated by the D_{st}-interpretation.

We also have:

Proposition 5.35 ([14]). *The system $\mathsf{H} :\equiv \mathsf{E\text{-}HA}^{\omega *}_{\mathrm{st}} + \mathrm{I} + \mathsf{NCR} + \mathsf{HAC} + \mathsf{HGMP}^{\mathrm{st}} + \mathsf{HIP}_{\forall^{\mathrm{st}}}$ is closed under both transfer rules, $\mathsf{TR}_{\forall}$ and $\mathsf{TR}_{\exists}$.*

5.4 The system E-PA$^{\omega*}_{\mathrm{st}}$ and negative translation

By combining the functional interpretation from the previous section with negative translation we can obtain conservation and term extraction results for classical systems as well. We will work out the details in this and the next section.

First, we need to set up a suitable classical system E-PA$^{\omega*}_{\mathrm{st}}$. It will be an extension of E-PA$^{\omega*}$, which is E-HA$^{\omega*}$ with the law of excluded middle added for all formulas. When working with classical systems, we will often take the logical connectives $\neg, \vee, \forall$ as primitive and regard the others as defined. In a similar spirit, the language of E-PA$^{\omega*}_{\mathrm{st}}$ will be that of E-PA$^{\omega*}$ extended just with unary predicates st^σ for every type $\sigma \in T^*$; the external quantifiers $\forall^{\mathrm{st}}, \exists^{\mathrm{st}}$ are regarded as abbreviations:

$$\begin{aligned} \forall^{\mathrm{st}} x\, \Phi(x) &:\equiv \forall x(\, \mathrm{st}(x) \to \Phi(x)\,), \\ \exists^{\mathrm{st}} x\, \Phi(x) &:\equiv \exists x(\, \mathrm{st}(x) \wedge \Phi(x)\,). \end{aligned}$$

Definition 5.36 (E-PA$^{\omega*}_{\mathrm{st}}$). The system E-PA$^{\omega*}_{\mathrm{st}}$ is

$$\text{E-PA}^{\omega*}_{\mathrm{st}} := \text{E-PA}^{\omega*} + \mathcal{T}^*_{\mathrm{st}} + \mathsf{IA}^{\mathrm{st}}$$

where

- $\mathcal{T}^*_{\mathrm{st}}$ consists of:
  1. the schema $\mathrm{st}(x) \wedge x = y \to \mathrm{st}(y)$,
  2. a schema providing for each closed term t in $\mathcal{T}^*$ the axiom $\mathrm{st}(t)$,
  3. the schema $\mathrm{st}(f) \wedge \mathrm{st}(x) \to \mathrm{st}(fx)$.
- $\mathsf{IA}^{\mathrm{st}}$ is the external induction axiom:

$$\mathsf{IA}^{\mathrm{st}} \quad : \quad (\Phi(0) \wedge \forall^{\mathrm{st}} n^0 (\Phi(n) \to \Phi(n+1))) \to \forall^{\mathrm{st}} n^0 \Phi(n).$$

Again we warn the reader that the induction axiom from E-PA$^{\omega*}$

$$(\varphi(0) \wedge \forall n^0 (\varphi(n) \to \varphi(n+1))) \to \forall n^0 \varphi(n)$$

is supposed to apply to internal formulas φ only.

As for E-HA$^{\omega*}_{\mathrm{st}}$, we have:

Proposition 5.37. *If a formula Φ is provable in E-PA$^{\omega*}_{\mathrm{st}}$, then its internalization Φ^{int} is provable in E-PA$^{\omega*}$. Hence E-PA$^{\omega*}_{\mathrm{st}}$ is a conservative extension of E-PA$^{\omega*}$ and E-PA$^{\omega}$.*

We will now show how negative translation provides an interpretation of E-PA$^{\omega*}_{\mathrm{st}}$ in E-HA$^{\omega*}_{\mathrm{st}}$. Various negative translations exist, with the one due to Gödel and Gentzen being the most well-known. Here, we work with two variants, the first of which is due to Kuroda [91], see also section 1.5.

Definition 5.38 (Kuroda's negative translation for E-PA$^{\omega*}_{\mathrm{st}}$). For an arbitrary formula Φ in the language of E-PA$^{\omega}_{\mathrm{st}}$, we define its Kuroda negative translation in E-HA$^{\omega*}_{\mathrm{st}}$ as

$$\Phi^{\mathrm{Ku}} :\equiv \neg\neg\Phi_{\mathrm{Ku}},$$

where Φ_{Ku} is defined inductively on the structure of Φ as follows:

$$\begin{aligned} \Phi_{\mathsf{Ku}} &:\equiv \Phi \quad \text{for atomic formulas } \Phi, \\ (\neg\Phi)_{\mathsf{Ku}} &:\equiv \neg\Phi_{\mathsf{Ku}}, \\ (\Phi \vee \Psi)_{\mathsf{Ku}} &:\equiv \Phi_{\mathsf{Ku}} \vee \Psi_{\mathsf{Ku}}, \\ (\forall x\, \Phi(x))_{\mathsf{Ku}} &:\equiv \forall x\, \neg\neg\Phi_{\mathsf{Ku}}(x). \end{aligned}$$

Theorem 5.39. $\mathsf{E\text{-}PA}^{\omega*}_{\mathrm{st}} \vdash \Phi \leftrightarrow \Phi^{\mathsf{Ku}}$ *and if* $\mathsf{E\text{-}PA}^{\omega*}_{\mathrm{st}} + \Delta \vdash \Phi$ *then* $\mathsf{E\text{-}HA}^{\omega*}_{\mathrm{st}} + \Delta^{\mathsf{Ku}} \vdash \Phi^{\mathsf{Ku}}$.

Proof. It is clear that, classically, Φ, Φ_{Ku} and Φ^{Ku} are all equivalent. The second statement is proved by induction on the proof of $\mathsf{E\text{-}PA}^{\omega*}_{\mathrm{st}} + \Delta \vdash \Phi$. For the cases of the axioms and rules of classical logic and $\mathsf{E\text{-}PA}^{\omega*}$, see, for instance, [71, Proposition 10.3]. As the Kuroda negative translation of every instance of $\mathcal{T}^*_{\mathrm{st}}$ or $\mathsf{IA}^{\mathrm{st}}$ is provable in $\mathsf{E\text{-}HA}^{\omega*}_{\mathrm{st}}$, the statement is proved. □

It will turn out to be convenient to introduce a second negative translation, extracted from the work of Krivine by Streicher and Reus (see [89, 119, 118]). This translation will interpret $\mathsf{E\text{-}PA}^{\omega*}_{\mathrm{st}}$ into $\mathsf{E\text{-}HA}^{\omega*}_{\mathrm{nst}}$ (see Remark 5.31).

Definition 5.40 (Krivine's negative translation for $\mathsf{E\text{-}PA}^{\omega*}_{\mathrm{st}}$)**.** For an arbitrary formula Φ in the language of $\mathsf{E\text{-}PA}^{\omega*}_{\mathrm{st}}$, we define its Krivine negative translation in $\mathsf{E\text{-}HA}^{\omega*}_{\mathrm{nst}}$ as

$$\Phi^{\mathsf{Kr}} :\equiv \neg\Phi_{\mathsf{Kr}},$$

where Φ_{Kr} is defined inductively on the structure of Φ as follows

$$\begin{aligned} \varphi_{\mathsf{Kr}} &:\equiv \neg\varphi \quad \text{for an internal atomic formula } \varphi, \\ \mathrm{st}(x)_{\mathsf{Kr}} &:\equiv \mathrm{nst}(x), \\ (\neg\Phi)_{\mathsf{Kr}} &:\equiv \neg\Phi_{\mathsf{Kr}}, \\ (\Phi \vee \Psi)_{\mathsf{Kr}} &:\equiv \Phi_{\mathsf{Kr}} \wedge \Psi_{\mathsf{Kr}}, \\ (\forall x\, \Phi(x))_{\mathsf{Kr}} &:\equiv \exists x\, \Phi_{\mathsf{Kr}}(x). \end{aligned}$$

Theorem 5.41. *For every formula* Φ *in the language of* $\mathsf{E\text{-}PA}^{\omega*}_{\mathrm{st}}$*, we have:*

1. $\mathsf{E\text{-}HA}^{\omega*}_{\mathrm{nst}} \vdash \Phi^{\mathsf{Kr}} \leftrightarrow \Phi^{\mathsf{Ku}}$.
2. *If* $\mathsf{E\text{-}PA}^{\omega*}_{\mathrm{st}} + \Delta \vdash \Phi$*, then* $\mathsf{E\text{-}HA}^{\omega*}_{\mathrm{nst}} + \Delta^{\mathsf{Kr}} \vdash \Phi^{\mathsf{Kr}}$.

Proof. Item 1 is easily proved by induction on the structure of Φ. Item 2 follows from item 1 and Theorem 5.39. □

5.5 A functional interpretation for $\mathsf{E\text{-}PA}^{\omega*}_{\mathrm{st}}$

We will now combine negative translation and our functional interpretation D_{st} to obtain a functional interpretation of the classical system $\mathsf{E\text{-}PA}^{\omega*}_{\mathrm{st}}$.

The interpretation

Definition 5.42. (S_{st}-interpretation for E-PA$^{\omega*}_{\mathrm{st}}$.) To each formula $\Phi(\underline{a})$ with free variables $\underline{a}$ in the language of E-PA$^{\omega*}_{\mathrm{st}}$ we associate its S_{st}-interpretation

$$\Phi^{S_{\mathrm{st}}}(\underline{a}) :\equiv \forall^{\mathrm{st}}\underline{x}\,\exists^{\mathrm{st}}\underline{y}\,\varphi_S(\underline{x},\underline{y},\underline{a}),$$

where φ_S is an internal formula. Moreover, $\underline{x}$ and $\underline{y}$ are tuples of variables whose length and types depend only on the logical structure of Φ. The interpretation of the formula is defined inductively on its structure. If

$$\Phi^{S_{\mathrm{st}}}(\underline{a}) :\equiv \forall^{\mathrm{st}}\underline{x}\,\exists^{\mathrm{st}}\underline{y}\;\varphi_S(\underline{x},\underline{y},\underline{a}) \text{ and } \Psi^{S_{\mathrm{st}}}(\underline{b}) :\equiv \forall^{\mathrm{st}}\underline{u}\,\exists^{\mathrm{st}}\underline{v}\;\psi_S(\underline{u},\underline{v},\underline{b}),$$

then

(i) $\varphi^{S_{\mathrm{st}}} :\equiv \varphi$ for atomic internal $\varphi(\underline{a})$,

(ii) $\big(\mathrm{st}(z)\big)^{S_{\mathrm{st}}} :\equiv \exists^{\mathrm{st}}x\,(z = x)$,

(iii) $(\neg\Phi)^{S_{\mathrm{st}}} :\equiv \forall^{\mathrm{st}}\underline{Y}\exists^{\mathrm{st}}\underline{x}\,\forall\underline{y}\in\underline{Y}[\underline{x}]\neg\varphi_S(\underline{x},\underline{y},\underline{a})$,

(iv) $(\Phi\vee\Psi)^{S_{\mathrm{st}}} :\equiv \forall^{\mathrm{st}}\underline{x},\underline{u}\,\exists^{\mathrm{st}}\underline{y},\underline{v}\,\big(\varphi_S(\underline{x},\underline{y},\underline{a})\vee\psi_S(\underline{u},\underline{v},\underline{b})\big)$,

(v) $(\forall z\,\varphi)^{S_{\mathrm{st}}} :\equiv \forall^{\mathrm{st}}\underline{x}\exists^{\mathrm{st}}\underline{y}\forall z\exists\underline{y}'\in\underline{y}\,\varphi_S(\underline{x},\underline{y}',z)$.

Theorem 5.43. (Soundness of the S_{st}-interpretation.) *Let $\Phi(\underline{a})$ be a formula in the language of* E-PA$^{\omega*}_{\mathrm{st}}$ *and suppose $\Phi(\underline{a})^{S_{\mathrm{st}}} \equiv \forall^{\mathrm{st}}\underline{x}\,\exists^{\mathrm{st}}\underline{y}\,\varphi(\underline{x},\underline{y},\underline{a})$. If Δ_{int} is a collection of internal formulas and*

$$\text{E-PA}^{\omega*}_{\mathrm{st}} + \Delta_{\mathrm{int}} \vdash \Phi(\underline{a}),$$

then one can extract from the formal proof a sequence of closed terms $\underline{t}$ in $\mathcal{T}^*$ such that

$$\text{E-PA}^{\omega*} + \Delta_{\mathrm{int}} \vdash \forall\underline{x}\exists\underline{y}\in\underline{t}(\underline{x})\;\varphi(\underline{x},\underline{y},\underline{a}).$$

Our proof of this theorem relies on the following lemma:

Lemma 5.44. *Let $\Phi(\underline{a})$ be a formula in the language of* E-PA$^{\omega*}_{\mathrm{st}}$ *and assume*

$$\begin{aligned}\Phi^{S_{\mathrm{st}}} &\equiv \forall^{\mathrm{st}}\underline{x}\exists^{\mathrm{st}}\underline{y}\,\varphi(\underline{x},\underline{y},\underline{a}) \quad \textit{and}\\ (\Phi_{\mathrm{Kr}})^{D_{\mathrm{st}}} &\equiv \exists^{\mathrm{st}}\underline{u}\forall^{\mathrm{st}}\underline{v}\,\theta(\underline{u},\underline{v},\underline{a}).\end{aligned}$$

Then the tuples $\underline{x}$ and $\underline{u}$ have the same length and the variables they contain have the same types. The same applies to $\underline{y}$ and $\underline{v}$. In addition, we have

$$\text{E-PA}^{\omega*} \vdash \varphi(\underline{x},\underline{y},\underline{a}) \leftrightarrow \neg\theta(\underline{x},\underline{y},\underline{a}).$$

Proof. The proof is by induction on the structure of Φ.

(i) If $\Phi \equiv \psi$, an internal and atomic formula, then $\varphi \equiv \psi$ and $\theta \equiv \neg\psi$, so E-PA$^{\omega*}$ $\vdash$ $\varphi \leftrightarrow \neg\theta$.

(ii) If $\Phi \equiv \mathrm{st}(z)$, then $\varphi \equiv y = z$ and $\theta \equiv y \neq z$, so E-PA$^{\omega*}$ $\vdash \varphi \leftrightarrow \neg\theta$.

(iii) If $\Phi \equiv \neg\Phi'$ with $(\Phi')^{S_{\text{st}}} \equiv \forall^{\text{st}}\underline{x}\exists^{\text{st}}\underline{y}\,\varphi'(\underline{x},\underline{y},\underline{a})$ and $(\Phi'_{\mathsf{Kr}})^{D_{\text{st}}} \equiv \exists^{\text{st}}\underline{u}\forall^{\text{st}}\underline{v}\,\theta'(\underline{u},\underline{v},\underline{a})$, then $\varphi \equiv \forall \underline{y}' \in \underline{Y}[\underline{x}]\, \neg\varphi'(\underline{x},\underline{y}')$ and $\theta \equiv \neg\forall \underline{i} \in \underline{Y}[\underline{x}]\, \theta'(\underline{x},\underline{i})$. Since $\mathsf{E\text{-}PA}^{\omega*} \vdash \varphi' \leftrightarrow \neg\theta'$ by induction hypothesis, also $\mathsf{E\text{-}PA}^{\omega*} \vdash \varphi \leftrightarrow \neg\theta$.

(iv) If $\Phi \equiv \Phi_0 \vee \Phi_1$ with

$$\Phi_i^{S_{\text{st}}} \equiv \forall^{\text{st}}\underline{x}\exists^{\text{st}}\underline{y}\,\varphi_i(\underline{x},\underline{y},\underline{a})$$

and

$$((\Phi_i)_{\mathsf{Kr}})^{D_{\text{st}}} \equiv \exists^{\text{st}}\underline{u}\forall^{\text{st}}\underline{v}\,\theta_i(\underline{u},\underline{v},\underline{a}),$$

then $\varphi \equiv \varphi_0 \vee \varphi_1$ and $\theta \equiv \theta_0 \wedge \theta_1$. Since $\mathsf{E\text{-}PA}^{\omega*} \vdash \varphi_i \leftrightarrow \neg\theta_i$ by induction hypothesis, also $\mathsf{E\text{-}PA}^{\omega*} \vdash \varphi \leftrightarrow \neg\theta$.

(v) If $\Phi \equiv \forall z\, \Phi'$ with

$$(\Phi')^{S_{\text{st}}} \equiv \forall^{\text{st}}\underline{x}\exists^{\text{st}}\underline{y}\,\varphi'(\underline{x},\underline{y},z,\underline{a})$$

and

$$(\Phi'_{\mathsf{Kr}})^{D_{\text{st}}} \equiv \exists^{\text{st}}\underline{u}\forall^{\text{st}}\underline{v}\,\theta'(\underline{u},\underline{v},z,\underline{a}),$$

then $\varphi \equiv \forall z\exists \underline{y}' \in \underline{y}\varphi'(\underline{x},\underline{y}',z,\underline{a})$ and $\theta \equiv \exists z\forall \underline{y}' \in \underline{y}\,\theta'(\underline{x},\underline{y}',z,\underline{a})$. Since $\mathsf{E\text{-}PA}^{\omega*} \vdash \varphi' \leftrightarrow \neg\theta'$ by induction hypothesis, also $\mathsf{E\text{-}PA}^{\omega*} \vdash \varphi \leftrightarrow \neg\theta$.

□

Remark. This lemma is the reason why we introduced the system $\mathsf{E\text{-}HA}^{\omega*}_{\text{nst}}$ in the Remark 5.31: it would fail if we would let the Krivine negative translation land directly in $\mathsf{E\text{-}HA}^{\omega*}_{\text{st}}$ with $\text{st}(z)_{\mathsf{Kr}} = \neg\,\text{st}(z)$. As it is, this lemma yields a quick proof of the soundness of the S_{st}-interpretation.

Proof. (Of the soundness of the S_{st}-interpretation, Theorem 5.43.) Let $\Phi(\underline{a})$ be a formula in the language of $\mathsf{E\text{-}PA}^{\omega*}_{\text{st}}$ and let φ and θ be such that

$$\begin{aligned} \Phi^{S_{\text{st}}} &\equiv \forall^{\text{st}}\underline{x}\exists^{\text{st}}\underline{y}\,\varphi(\underline{x},\underline{y},\underline{a}), \\ (\Phi_{\mathsf{Kr}})^{D_{\text{st}}} &\equiv \exists^{\text{st}}\underline{x}\forall^{\text{st}}\underline{y}\,\theta(\underline{x},\underline{y},\underline{a}) \end{aligned}$$

and $\mathsf{E\text{-}PA}^{\omega*} \vdash \varphi \leftrightarrow \neg\theta$, as in Lemma 5.44.

Now, suppose that Δ_{int} is a set of internal formulas and $\Phi(\underline{a})$ is a formula provable in $\mathsf{E\text{-}PA}^{\omega*}_{\text{st}}$ from Δ_{int}. We first apply soundness of the Krivine negative translation (Theorem 5.41) to see that

$$\mathsf{E\text{-}HA}^{\omega*}_{\text{nst}} + \Delta^{\mathsf{Kr}}_{\text{int}} \vdash \Phi^{\mathsf{Kr}},$$

where $\Phi^{\mathsf{Kr}} \equiv \neg\Phi_{\mathsf{Kr}}$. So if $(\Phi_{\mathsf{Kr}})^{D_{\text{st}}} \equiv \exists^{\text{st}}\underline{x}\forall^{\text{st}}\underline{y}\,\theta(\underline{x},\underline{y},\underline{a})$, then

$$(\Phi^{\mathsf{Kr}})^{D_{\text{st}}} \equiv \exists^{\text{st}}\underline{Y}\forall^{\text{st}}\underline{x}\exists\underline{y} \in \underline{Y}[\underline{x}]\neg\theta(\underline{x},\underline{y},\underline{a}).$$

It follows from the soundness theorem for D_{st} (Theorem 5.30) and Remark 5.31 that there is a sequence of closed terms $\underline{s}$ from $\mathcal{T}^*$ such that

$$\mathsf{E\text{-}HA}^{\omega*} + \Delta^{\mathsf{Kr}}_{\text{int}} \vdash \forall\underline{x}\exists\underline{y} \in \underline{s}[\underline{x}]\neg\theta(\underline{x},\underline{y},\underline{a}).$$

Since $\mathsf{E\text{-}PA}^{\omega*} \vdash \Delta^{\mathsf{Kr}}_{\text{int}} \leftrightarrow \Delta_{\text{int}}$ and $\mathsf{E\text{-}PA}^{\omega*} \vdash \varphi \leftrightarrow \neg\theta$ we have

$$\mathsf{E\text{-}PA}^{\omega*} + \Delta_{\text{int}} \vdash \forall\underline{x}\exists\underline{y} \in \underline{t}(\underline{x})\,\varphi(\underline{x},\underline{y},\underline{a}),$$

with $\underline{t} \equiv \lambda\underline{x}.\underline{s}[\underline{x}]$. □

Characteristic principles

The characteristic principles of our functional interpretation for classical arithmetic are idealization I (or, equivalently, R: see Section 4.1) and HAC_{int}

$$\forall^{\text{st}} x \exists^{\text{st}} y\, \varphi(x,y) \to \exists^{\text{st}} F \forall^{\text{st}} x \exists y \in F(x)\, \varphi(x,y),$$

which is the choice scheme HAC restricted to internal formulas. To see this, note first of all that we have:

Proposition 5.45. *For any formula Φ in the language of $\text{E-PA}^{\omega*}_{\text{st}}$ one has:*

$$\text{E-PA}^{\omega*}_{\text{st}} + \text{I} + \text{HAC}_{\text{int}} \vdash \Phi \leftrightarrow \Phi^{S_{\text{st}}}.$$

Proof. An easy proof by induction on the structure of Φ, using HAC_{int} for the case of negation and I (or rather R) in the case of internal universal quantification. □

For the purpose of showing that I and HAC_{int} are interpreted, it will be convenient to consider the "hybrid" system $\text{E-HA}^{\omega*}_{\text{nst}} + \text{LEM}_{\text{int}}$, where LEM_{int} is the law of excluded middle for internal formulas. For this hybrid system we have the following easy lemma, whose proof we omit:

Lemma 5.46. *We have:*

1. $\text{E-HA}^{\omega*}_{\text{nst}} + \text{LEM}_{\text{int}} \vdash \varphi^{\text{Ku}} \leftrightarrow \varphi$, *if φ is an internal formula in the the language of* $\text{E-PA}^{\omega*}_{\text{st}}$.
2. $\text{E-HA}^{\omega*}_{\text{nst}} + \text{LEM}_{\text{int}} + \text{I} \vdash \text{I}^{\text{Ku}}$.
3. $\text{E-HA}^{\omega*}_{\text{nst}} + \text{LEM}_{\text{int}} + \text{HAC}_{\text{int}} + \text{HGMP}^{\text{st}} \vdash \text{HAC}^{\text{Ku}}_{\text{int}}$.

This means we can strengthen Theorem 5.43 to:

Theorem 5.47. (Soundness of the S_{st}-interpretation, full version.) *Let $\Phi(\underline{a})$ be a formula in the language of $\text{E-PA}^{\omega*}_{\text{st}}$ and suppose $\Phi(\underline{a})^{S_{\text{st}}} \equiv \forall^{\text{st}} \underline{x}\, \exists^{\text{st}} \underline{y}\, \varphi(\underline{x}, \underline{y}, \underline{a})$. If Δ_{int} is a collection of internal formulas and*

$$\text{E-PA}^{\omega*}_{\text{st}} + \text{I} + \text{HAC}_{\text{int}} + \Delta_{\text{int}} \vdash \Phi(\underline{a}),$$

then one can extract from the formal proof a sequence of closed terms $\underline{t}$ in $\mathcal{T}^*$ such that

$$\text{E-PA}^{\omega*} + \Delta_{\text{int}} \vdash \forall \underline{x} \exists \underline{y} \in \underline{t}(\underline{x})\, \varphi(\underline{x}, \underline{y}, \underline{a}).$$

Proof. The argument is a slight extension of the proof of Theorem 5.43. So, once again, let $\Phi(\underline{a})$ be a formula in the language of $\text{E-PA}^{\omega*}_{\text{st}}$ and φ and θ be such that

$$\begin{aligned} \Phi^{S_{\text{st}}} &\equiv \forall^{\text{st}} \underline{x} \exists^{\text{st}} \underline{y}\, \varphi(\underline{x}, \underline{y}, \underline{a}), \\ (\Phi_{\text{Kr}})^{D_{\text{st}}} &\equiv \exists^{\text{st}} \underline{x} \forall^{\text{st}} \underline{y}\, \theta(\underline{x}, \underline{y}, \underline{a}) \end{aligned}$$

and $\text{E-PA}^{\omega*} \vdash \varphi \leftrightarrow \neg\theta$, as in Lemma 5.44.

This time we suppose Δ_{int} is a set of internal formulas and $\Phi(\underline{a})$ is a formula provable in $\text{E-PA}^{\omega*}_{\text{st}}$ from $\text{I} + \text{HAC}_{\text{int}} + \Delta_{\text{int}}$. We first apply soundness of the Kuroda negative translation (Theorem 5.39), which yields:

$$\text{E-HA}^{\omega*}_{\text{nst}} + \text{I}^{\text{Ku}} + \text{HAC}^{\text{Ku}}_{\text{int}} + \Delta^{\text{Ku}}_{\text{int}} \vdash \Phi^{\text{Ku}}.$$

Then the previous lemma implies that:

$$\text{E-HA}^{\omega*}_{\text{nst}} + \text{LEM}_{\text{int}} + \text{I} + \text{HAC}_{\text{int}} + \text{HGMP}^{\text{st}} + \Delta^{\text{Ku}}_{\text{int}} \vdash \Phi^{\text{Ku}}.$$

Note that $\text{E-HA}^{\omega*}_{\text{nst}} \vdash \Phi^{\text{Ku}} \leftrightarrow \Phi^{\text{Kr}}$, $\Phi^{\text{Kr}} \equiv \neg\Phi_{\text{Kr}}$ and

$$(\Phi^{\text{Kr}})^{D_{\text{st}}} \equiv \exists^{\text{st}}\underline{Y}\forall^{\text{st}}\underline{x}\exists\underline{y} \in \underline{Y}[\underline{x}]\neg\theta(\underline{x},\underline{y},\underline{a}).$$

Therefore the soundness theorem for D_{st} (Theorem 5.30), in combination with Remark 5.31 and the fact that the axiom scheme LEM_{int} is internal, implies that there is a sequence of closed terms $\underline{s}$ from $\mathcal{T}^*$ such that

$$\text{E-HA}^{\omega*} + \text{LEM} + \Delta^{\text{Ku}}_{\text{int}} \vdash \forall\underline{x}\exists\underline{y} \in \underline{s}[\underline{x}]\neg\theta(\underline{x},\underline{y},\underline{a}).$$

Since $\text{E-PA}^{\omega*} \vdash \text{LEM}$, $\text{E-PA}^{\omega*} \vdash \Delta^{\text{Ku}}_{\text{int}} \leftrightarrow \Delta_{\text{int}}$ and $\text{E-PA}^{\omega*} \vdash \varphi \leftrightarrow \neg\theta$, we have

$$\text{E-PA}^{\omega*} + \Delta_{\text{int}} \vdash \forall\underline{x}\exists\underline{y} \in \underline{t}(\underline{x})\ \varphi(\underline{x},\underline{y},\underline{a})$$

with $\underline{t} \equiv \lambda\underline{x}.\underline{s}[\underline{x}]$. □

The following picture depicts the relation between the various interpretations we have established:

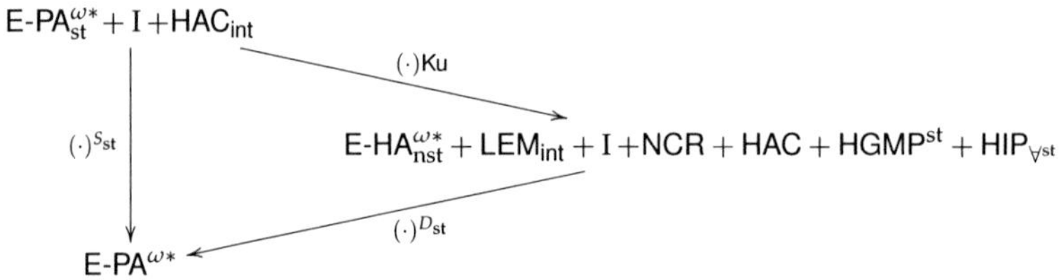

Figure 2: The Shoenfield and negative Dialectica interpretations.

Conservation results and the transfer principle

Theorem 5.47 immediately gives us the following conservation result:

Corollary 5.48. $\text{E-PA}^{\omega*}_{\text{st}} + \text{I} + \text{HAC}_{\text{int}}$ *is a conservative extension of* $\text{E-PA}^{\omega*}$ *and hence of* E-PA^{ω}.

Interpreting DNS for standard arguments

We already discussed that solving the negative interpretation of DNS and thereby that of choice and comprehension was of major interest to proof mining in the introduction and implicitly at several occasions in the first chapter.
Therefore it seems natural to have a closer look at the corresponding scenario in our non-standard setting, namely the S_{st}-interpretation of the DNSst principle. Recall:

$$\text{DNSst} \quad : \quad \forall^{\text{st}}n^0\,\exists y^\tau\,\Phi(n,y) \to \exists f^{0\to\tau}\,\forall^{\text{st}}n^0\,\Phi(n,f(n)).$$

It turns out that the interpretation (obtained with the help of an automated interpreter due to myself) remains very similar to the standard scenario and reads as follows:

$$
\begin{aligned}
&\exists^{\mathrm{st}}\hat{U},\hat{Y},\tilde{N}\forall^{\mathrm{st}}\hat{V},N,\hat{X}(\\
&\quad \forall\hat{n}\leq|\hat{Y}[\hat{X},\hat{V},N]|,\hat{p}\leq|\tilde{N}[\hat{X},\hat{V},N]|\\
&\quad \neg\forall\hat{b}\leq|\hat{X}[\tilde{N}[\hat{X},\hat{V},N]_{\hat{p}}][\hat{Y}[\hat{X},\hat{V},N]_{\hat{n}}]|\\
&\quad \neg\forall\hat{a}\leq|\hat{Y}[\hat{X},\hat{V},N]_{\hat{n}}[\hat{X}[\tilde{N}[\hat{X},\hat{V},N]_{\hat{p}}][\hat{Y}[\hat{X},\hat{V},N]_{\hat{n}}]_{\hat{b}}]|\\
&\qquad \varphi(\tilde{N}[\hat{X},\hat{V},N]_{\hat{p}},\hat{X}[\tilde{N}[\hat{X},\hat{V},N]_{\hat{p}}][\hat{Y}[\hat{X},\hat{V},N]_{\hat{n}}]_{\hat{b}},\hat{Y}[\hat{X},\hat{V},N]_{\hat{n}}[\hat{X}[\tilde{N}[\hat{X},\hat{V},N]_{\hat{p}}][\hat{Y}[\hat{X},\hat{V},N]_{\hat{n}}]_{\hat{b}}]_{\hat{a}})\rightarrow\\
&\quad \neg\forall\hat{m}\leq|\hat{U}[\hat{X}][\hat{V},N]|\\
&\quad \neg\forall\hat{k}\leq|\hat{V}[\hat{U}[\hat{X}][\hat{V},N]_{\hat{m}}]|,\hat{l}\leq|N[\hat{U}[\hat{X}][\hat{V},N]_{\hat{m}}]|\\
&\qquad \varphi(N[\hat{U}[\hat{X}][\hat{V},N]_{\hat{m}}]_{\hat{l}},\hat{U}[\hat{X}][\hat{V},N]_{\hat{m}}[N[\hat{U}[\hat{X}][\hat{V},N]_{\hat{m}}]_{\hat{l}}],\hat{V}[\hat{U}[\hat{X}][\hat{V},N]_{\hat{m}}]_{\hat{k}})).
\end{aligned}
$$

In particular, note that the equations which need to be solved correspond one to one to the standard case (see 1.13 or [71]), except that here, of course, we have to use our sequence application Λ. In fact, to obtain a solution we might need to define a modified version of the bar recursor, similarly as we did for normal recursion $\underline{\mathcal{R}}_{\underline{\rho}}$.

5.6 Countable saturation

The main idea behind providing a functional interpretation for nonstandard proofs, is that we would like to see if the interpretation that we have developed in this chapter could be used to "unwind" or "proof-mine" nonstandard arguments. Nonstandard arguments have been used in areas where proof-mining techniques have also been successful, such as metric fixed point theory (for methods of nonstandard analysis applied to metric fixed point theory, see [2, 58]; for application of proof-mining to metric fixed point theory, see [20, 32, 69, 79, 81, 92]) and ergodic theory (for a nonstandard proof of an ergodic theorem, see [55]; for applications of proof-mining to ergodic theory, see [6, 8, 34, 35, 73, 80, 107]), therefore this looks quite promising. For the former type of applications to work in full generality, one would have to extend our functional interpretation to include types for abstract metric spaces, as in [37, 70].

Towards that purpose, one should extend our methods to allow one to prove conservativity results over $\mathsf{WE\text{-}HA}^\omega$ and $\mathsf{WE\text{-}PA}^\omega$ as well: this will be important if one wishes to combine the results presented here with the proof-mining techniques from [71].

Most importantly for now, we would like to understand the use of saturation principles in nonstandard arguments. These are of particular interest since they are used in the construction of Loeb measures, which belong to one of the most successful nonstandard techniques. The general saturation principle is

$$\mathsf{SAT}: \quad \forall^{\mathrm{st}}x^\sigma\,\exists y^\tau\,\Phi(x,y)\rightarrow\exists f^{\sigma\rightarrow\tau}\,\forall^{\mathrm{st}}x^\sigma\,\Phi(x,f(x)).$$

Whether this principle has a D_{st}-interpretation within Gödel's $\mathcal{T}^*$, we do not know; but Countable Saturation principle

$$\mathsf{CSAT}: \quad \forall^{\mathrm{st}}n^0\,\exists y^\tau\,\Phi(n,y)\rightarrow\exists f^{0\rightarrow\tau}\,\forall^{\mathrm{st}}n^0\,\Phi(n,f(n))$$

has and that seems to be sufficient for the construction of Loeb measures. However, as in the standard scenario, the more interesting context here is the classical one. Interpreting CSAT and SAT in using the S_{st}-interpretation is actually quite difficult and it is most probable that the analysis will require some form of bar recursion.
Actually, as we can see already on the formalization of CSAT above, it turns out that the situation is very similar as discussed in section 1.14 regarding AC.
We recall:

$$\begin{aligned}
\mathrm{I} &: \quad \forall^{st} x^\sigma \exists y^\tau \forall i \leq |x|\, \varphi((x)_i, y) \rightarrow \exists y^\tau \forall^{st} x^\sigma\, \varphi(x,y), \\
\mathsf{HAC}_{\mathsf{int}} &: \quad \forall^{st} x \exists^{st} y\, \varphi(x,y) \rightarrow \exists^{st} F \forall^{st} x \exists i \leq |F(x)|\, \varphi(x, F(x)_i), \quad F(x)_i :\equiv (F(x))_i, \\
\mathsf{AC}_0^{st} &: \quad \forall^{st} n^0 \exists^{st} x \Phi(n,x) \ \rightarrow \exists^{st} f \forall^{st} n^0 \Phi(n, f(n)), \\
\mathsf{CSAT} &: \quad \forall^{st} n^0\, \exists y^\tau\, \Phi(n,y) \rightarrow \exists f^{0\rightarrow\tau}\, \forall^{st} n^0\, \Phi(n, f(n))
\end{aligned}$$

Theorem 5.49 (Main claim).

$$\mathsf{E\text{-}PA}^\omega_{st} + \mathrm{I} + \mathsf{HAC}_{\mathsf{int}} + \mathsf{AC}_0^{st} \ \vdash\ \mathsf{CSAT}.$$

Proof. Assuming

$$\mathsf{E\text{-}PA}^\omega_{st} + \mathrm{I} + \mathsf{HAC}_{\mathsf{int}} \ \vdash\ \Phi(x,y) \leftrightarrow \forall^{st} u \exists^{st} v\, \varphi(u,v,x,y),$$

(which follows from Proposition 8.5 in Berg et al.) it suffices to show that

$$\forall^{st} x^0 \exists y \forall^{st} u \exists^{st} v\, \varphi(u,v,x,y) \tag{10}$$

implies

$$\exists \tilde{y}^{0\rightarrow\tau} \forall^{st} x^0, u \exists^{st} v\, \varphi(u,v,x,\tilde{y}(x)). \tag{11}$$

By $\mathsf{HAC}_{\mathsf{int}}$ we get that (10) implies

$$\forall^{st} x^0 \exists y \exists^{st} V \forall^{st} u \exists i \leq |V(u)|\, \varphi(u, V(u)_i, x, y),$$

which is by IL equivalent to

$$\forall^{st} x^0 \exists^{st} V \exists y \forall^{st} u \exists i \leq |V(u)|\, \varphi(u, V(u)_i, x, y),$$

and by AC_0^{st} this implies that

$$\exists^{st} V \forall^{st} x^0 \exists y \forall^{st} u \exists i \leq |V(x,u)|\, \varphi(u, V(x,u)_i, x, y). \tag{10'}$$

We can strengthen (11) to an under $\mathsf{HAC}_{\mathsf{int}}$ (on u) and AC_0^{st} (on x) equivalent statement

$$\exists \tilde{y} \exists^{st} V \forall^{st} x^0, u \exists i \leq |V(x,u)|\, \varphi(u, V(x,u)_i, x, \tilde{y}(x)),$$

which is by IL equivalent to

$$\exists^{st} V \exists \tilde{y} \forall^{st} x^0, u \exists i \leq |V(x,u)|\, \varphi(u, V(x,u)_i, x, \tilde{y}(x)).$$

By I, this follows from

$$\exists^{st} V \forall^{st} x^0, u \exists \tilde{y}\, \forall j \leq |x|, k \leq |u|\, \exists i \leq |V(x_j, u_k)| \quad \varphi(u_k, V(x_j, u_k)_i, x_j, \tilde{y}(x_j)). \tag{11'}$$

Hence it suffices to show that (10′) implies (11′).
Now to do so, let some standard V satisfy (10′), i.e. we have

$$\forall^{\mathrm{st}} x^0 \exists y \forall^{\mathrm{st}} u \exists i \leq |V(x,u)| \; \varphi(u, V(x,u)_i, x, y), \tag{12}$$

and fix an arbitrary but standard x of type 0. Then we have for some finite (standard) natural number n that

$$x = \langle x_1, \ldots, x_n \rangle,$$

with x_i^0 standard. So by (12) we have for each $x_j \in \{x_1, \ldots, x_n\}$ an y_j s.t.

$$\forall^{\mathrm{st}} u \exists i \leq |V(x_j,u)| \; \varphi(u, V(x_j,u)_i, x_j, y_j),$$

in other words we obtain (simply consider $y = \langle y_1, \ldots, y_n \rangle$)

$$\exists y \forall^{\mathrm{st}} u \forall j \leq |x| \exists i \leq |V(x_j,u)| \; \varphi(u, V(x_j,u)_i, x_j, y_j).$$

Finally, let $\tilde{y}$ be the function $x_j \mapsto y_j$, (for $1 \leq j \leq n$), and constant 0 otherwise. [30] Then we have that

$$\forall^{\mathrm{st}} u \forall j \leq |x| \exists i \leq |V(x_j,u)| \; \varphi(u, V(x_j,u)_i, x_j, \tilde{y}(x_j)),$$

so in particular also

$$\exists \tilde{y} \forall^{\mathrm{st}} u \forall j \leq |x| \exists i \leq |V(x_j,u)| \; \varphi(u, V(x_j,u)_i, x_j, \tilde{y}(x_j)),$$

and hence

$$\forall^{\mathrm{st}} u \exists \tilde{y} \; \forall j \leq |x|, k \leq |u| \; \exists i \leq |V(x_j,u)| \; \varphi(u, V(x_j,u)_i, x_j, \tilde{y}(x_j)),$$

and therefore also (11′). □

[30] Note that such a $\tilde{y}$ is primitive recursively definable in x and y, in Particular it is a term in $\mathcal{T}_0$ with parameters x and y. However, it is not necessarily standard, as the parameter y – though its size is finite – does not need to code only standard elements and, therefore, does not need to be standard itself.

Zusammenfassung

Nach der Einführung von elementaren Definitionen und Techniken im ersten Kapitel (das auf [106] und [108] basiert) des sogenannten Proof Mining (siehe vor allem [71]) haben wir uns mit Beweisen von Cauchy-Aussagen befasst (siehe Kapitel 2 und [83]). Basierend auf [83] haben wir eine strikte Hierarchie von 4 Arten von finitärem Informationsgehalt (im Sinne des Fields-Medaillisten Terence Tao, siehe [123]) definiert, trennende Beispiele vorgestellt und mittels des neu definierten Begriffs der effektiven Lernbarkeit allgemeine logische Bedingungen angegeben, unter welchen die jeweilige Informationsart (bei entsprechender Analyse mittels der Techniken aus Kapitel 1) zu erwarten ist.
Dabei ist besonders interessant, dass diese Resultate zum ersten Mal die sehr spezifische Struktur der in der Praxis üblicherweise vorkommenden Schranken der Metastabilitätsraten (wieder im Sinne von Tao, siehe [122]) erklären. Darüberhinaus finden wir die Struktur aus unserem sehr einfachen Beispiel (in gewissem Sinne das einfachstmögliche) einer konvergenten Folge, die nicht lernbar ist, mit erstaunlicher Ähnlichkeit bei der Metastabilitätsschranke eines starken nichtlinearen Ergodentheorems von R. Wittmann (siehe [129]) wieder, die wir als eine weitere Anwendung von Proof Mining in Kapital 3 (siehe auch [107]) extrahieren.
Aus Sicht der Berechenbarkeitskomplexität ist allerdings auch diese Schranke eher einfach. Eines der zentralen Prinzipien, die häufig im Kern eines Konvergenzbeweises stehen, ist das Bolzano-Weierstraß-Theorem (siehe Definition 4.1). Diese Analyse wurde im Kapitel 4 wiedergegeben, das auf [106] und [108] basiert. Man sollte bemerken, dass kurz nach [108] dieses Theorem auch von anderen Forschungsgruppen auf finitären Gehalt mit ähnlichen Ergebnissen untersucht worden ist (vor allem in Bezug auf konkrete Instanzen bei Brattka et al. [17] und auf den genauen Zusammenhang mit Bar-Rekursion bei Oliva et al. [100] basierend auf [108]).
Dank der Techniken, die wir im Kapitel 4 anwenden (mit der Vorarbeit aus Kapitel 1), können wir im Kapitel 5 auch zum Proof Mining selbst beitragen. Wir bauen auf der Basis [14] eines längeren Projekts auf, das als Ziel hat, Proof Mining auf Beweise der Nichtstandardanalysis zu erweitern. Wie schon in [14] bemerkt, wäre einer der nächsten großen Schritte, eine Lösung der Shoenfield-Interpretation (siehe 5.5) von CSAT (Countable Saturation principle, siehe 5.6) anzugeben, da dies benötigt wird, um Beweise, die auf Loeb-Maßen basieren, zu führen. Zwar erreichen wir dieses ambitionierte Ziel nicht, dennoch können wir einen erfolgreich abgeschlossenen ersten Schritt im letzten Abschnitt des 5. Kapitels präsentieren.
Abgesehen von CSAT (siehe 5.6) und der praktischen Anwendung der Resultate aus [14] bietet sich die Aussicht an, weitere Aspekte des in [14] entwickelten Systems zu untersuchen. Zum Beispiel wie schwer es ist, WKL zu interpretieren, und was die richtige Formalisierung dieses Prinzips ist (siehe vergleichsweise [121, 116, 57]) oder wie man das Standardisierungsprinzip interpretiert.

References

[1] Y. Akama, S. Berardi, S. Hayashi, and U. Kohlenbach. An arithmetical hierarchy of the law of excluded middle and related principles. In *Proc. of the 19th Annual IEEE Symposium on Logic in Computer Science (LICS'04)*, pages 192–201. IEEE PressIEEE Press, 2004.

[2] A. Aksoy and M. Khamsi. *Nonstandard methods in fixed point theory*. Universitext. Springer-Verlag, New York, 1990.

[3] F. Aschieri. *Learning based on realizability for HA+EM1 and 1-Backtracking games: Soundness and completeness*. To appear in Ann. Pure Appl. Logic.

[4] F. Aschieri. A constructive analysis of learning in peano arithmetic. *Ann. Pure Appl. Logic*, 163(11):1448–1470, 2012.

[5] F. Aschieri and S. Berardi. A new use of friedman's translation: interactive realizability. In B. et als, editor, *Logic, Construction, Computation*. Ontos-Verlag Series in Mathematical Logic, 2012.

[6] J. Avigad. The metamathematics of ergodic theory. *Ann. Pure Appl. Logic*, 157(2-3):64–76, 2009.

[7] J. Avigad and S. Feferman. Gödels's functional ("Dialectica") interpretation *in s.* R. Buss (Editor): Handbook of proof theory. *Studies in Logic and the foundations of mathematics*, 137:337–405, 1998.

[8] J. Avigad, P. Gerhardy, and H. Towsner. Local stability of ergodic averages. *Trans. Amer. Math. Soc.*, 362(1):261–288, 2010.

[9] J. Avigad and J. Helzner. Transfer principles in nonstandard intuitionistic arithmetic. *Arch. Math. Logic*, 41(6):581–602, 2002.

[10] J. Avigad and J. Rute. Oscillation and the mean ergodic theorem for uniformly convex Banach spaces. *ArXiv e-prints (1203.4124)*, to appear in: Ergodic Theory and Dynamical Systems, Mar. 2012.

[11] J. B. Baillon. Un théorème de type ergodique pour les contractions non linéaires dans un espace de Hilbert. *C. R. Acad. Sci. Paris Sér. A-B*, 280(22):Aii, A1511–A1514, 1975.

[12] J. B. Baillon. Quelques propriétés de convergence asymptotique pour les contractions impaires. *C. R. Acad. Sci. Paris Sér. A-B*, 283(8):Aii, A587–A590, 1976.

[13] S. Berardi, T. Coquand, and S. Hayashi. Games with 1-backtracking. *Ann. Pure Appl. Logic*, 161:1254–1264, 2010.

[14] B. v. Berg, E. M. Briseid, and P. Safarik. A functional interpretation for nonstandard arithmetic. *Ann. Pure Appl. Logic*, 163(12):1962–1994, 2012.

[15] M. Bezem. Strongly Majorizable Functionals of Finite Type: A Model for Bar Recursion Containing Disonctinuous Functionals. *The Journal of Symbolic Logic*, 50:652–660, 1985.

[16] V. Brattka and G. Gherardi. Weihrauch degrees, omniscience principles and weak computability. *J. Symb. Log.*, 76(1):143–176, 2011.

[17] V. Brattka, G. Gherardi, and A. Marcone. The Bolzano-Weierstrass theorem is the jump of weak KHonig's lemma. *Ann. Pure Appl. Logic*, 163(6):623–655, 2012.

[18] H. Brézis and F. E. Browder. Nonlinear ergodic theorems. *Bull. Amer. Math. Soc.*, 82(6):959–961, 1976.

[19] E. M. Briseid. Proof Mining Applied to Fixed Point Theorems for Mappings of Contractive Type. *Master Thesis at Department of Mathematics at the University of Oslo*, 2005.

[20] E. M. Briseid. Logical aspects of rates of convergence in metric spaces. *J. Symbolic Logic*, 74(4):1401–1428, 2009.

[21] R. E. Bruck. A simple proof of the mean ergodic theorem for nonlinear contractions in Banach spaces. *Israel J. Math.*, 32(2-3):107–116, 1979.

[22] T. Coquand. A semantics of evidence for classical arithmetic. *J. Symbolic Logic*, 60:325–337, 1995.

[23] J. Diller and W. Nahm. Eine Variante zur Dialectica-Interpretation der Heyting-Arithmetik endlicher Typen. *Arch. Math. Logik Grundlagenforsch.*, 16:49–66, 1974.

[24] P. Engrácia. *Proof-theoretical studies on the bounded functional interpretation.* PhD thesis, University of Lisbon, 2009.

[25] S. Feferman. *Theories of finite type related to mathematical practice* in J. *Barwise (Editor): Handbook of Mathematical Logic.* North Holland, Amsterdam, 1977.

[26] F. Ferreira. Injecting uniformities into Peano arithmetic. *Ann. Pure Appl. Logic*, 157(2-3):122–129, 2009.

[27] F. Ferreira and P. Oliva. Bounded functional interpretation. *Ann. Pure Appl. Logic*, 135(1-3):73–112, 2005.

[28] H. Friedman. Classical and intuitionistically provably recursive functions. In S. D. Müller, G.H., editor, *Higher Set Theory*, pages 21–27. Springer LNM 669, 1978.

[29] J. Gaspar. Factorization of the Shoenfield-like bounded functional interpretation. *Notre Dame J. Form. Log.*, 50(1):53–60, 2009.

[30] J. Gaspar and U. Kohlenbach. On Tao's "finitary" infinite pigeonhole principle. *J. Symbolic Logic*, 75(1):355–371, 2010.

[31] A. Genel and J. Lindenstrauss. An example concerning fixed points. *Israel J. Math.*, 22(1):81–86, 1975.

[32] P. Gerhardy. A quantitative version of Kirk's fixed point theorem for asymptotic contractions. *J. Math. Anal. Appl.*, 316:339–345, 2006.

[33] P. Gerhardy. *Functional interpretation and modified realizability interpretation of the double-negation shift*. Preprint. BRICS, Department of Computer Science, University of Aarhus, Denmark, 2006.

[34] P. Gerhardy. Proof mining in topological dynamics. *Notre Dame J. Form. Log.*, 49(4):431–446, 2008.

[35] P. Gerhardy. Proof mining in practice. In *Logic Colloquium 2007*, volume 35 of *Lect. Notes Log.*, pages 82–91. Assoc. Symbol. Logic, La Jolla, CA, 2010.

[36] P. Gerhardy and U. Kohlenbach. Strongly uniform bounds from semi-constructive proofs. *Ann. Pure Appl. Logic*, 141:89–107, 2006.

[37] P. Gerhardy and U. Kohlenbach. General logical metatheorems for functional analysis. *Trans. Amer. Math. Soc.*, 360:2615–2660, 2008.

[38] K. Gödel. Zur intuitionistischen Arithmetik und Zahlentheorie. *Ergebnisse eines Mathematischen Kolloquiums*, 4:34–38, 1933.

[39] K. Gödel. Über eine bisher noch nicht benützte Erweiterung des finiten Standpunktes. *Dialectica*, 12:280–287, 1958.

[40] E. M. Gold. Language identification in the limit. *Information and Control*, 10:447–474, 1967.

[41] B. Halpern. Fixed points of nonexpanding maps. *Bull. Amer. Math. Soc.*, 73:957–961, 1967.

[42] S. Hayashi. Mathematics based on learning. In *Proc. 13th Internat. Conf. ALT 2002 Vol. 272*, pages 7–21. Springer Lecture Notes in Artificial Intelligence, 2002.

[43] S. Hayashi. Mathematics based on incremental learning – excluded middle and inductive inference original research article. *Theoretical Computer Science – Journal version of [42]*, 350:125–139, 2006.

[44] S. Hayashi and M. Nakata. Towards limit computable mathematics. In P. C. et al., editor, *TYPES 2000*, pages 125–144. Springer LNCS **2277**, 2002.

[45] C. Henson and H. Keisler. On the strength of nonstandard analysis. *J. Symbolic Logic*, 51(2):377–386, 1986.

[46] K. Higuchi and T. Kihara. Inside the muchnik degrees: discontinuity, learnability, and constructivism. Preprint, 2012.

[47] K. Higuchi and A. Pauly. The degree structure of weihrauch-reducibility. *Logical Methods in Computer Science*, 9(2), 2011.

[48] N. Hirano. Nonlinear ergodic theorems and weak convergence theorems. *J. Math. Soc. Japan*, 34(1):35–46, 1982.

[49] N. Hirano and W. Takahashi. Nonlinear ergodic theorems for nonexpansive mappings in Hilbert spaces. *Kodai Math. J.*, 2(1):11–25, 1979.

[50] W. A. Howard. Functional interpretation of bar induction by bar recursion. *Compositio Mathematica*, 20:107–124, 1968.

[51] W. A. Howard. Hereditarily majorizable functionals of finite type. In *A.S. Troelstra (Editor): Mathematical Investigation of Intuitionistic Arithmetic and Analysis*, pages 454–461. Springer-Varlag, Berlin · Heidelberg · New York, 1973.

[52] W. A. Howard. Ordinal analysis of terms of finite type. *The Journal of Symbolic Logic*, 45:493–504, 1980.

[53] W. A. Howard. Ordinal analysis of simple cases of bar recursion. *The Journal of Symbolic Logic*, 46:17–31, 1981.

[54] R. L. Jones, I. V. Ostrovskii, and J. M. Rosenblatt. Square functions in ergodic theory. *Ergodic Theory and Dynamical Systems*, 16:267–305, 1996.

[55] T. Kamae. A simple proof of the ergodic theorem using nonstandard analysis. *Israel J. Math.*, 42(4):284–290, 1982.

[56] H. Keisler. The strength of nonstandard analysis. In *The strength of nonstandard analysis*, pages 3–26. SpringerWienNewYork, Vienna, 2007.

[57] H. J. Keisler. Nonstandard arithmetic and recursive comprehension. *Ann. Pure Appl. Logic*, 161(8):1047–1062, 2010.

[58] W. A. Kirk. Fixed points of asymptotic contractions. *J. Math. Anal. Appl.*, 277(2):645–650, 2003.

[59] S. C. Kleene. *Introduction to mathematics.* North-Holland Publ. Co. (Amsterdam), P. Noordhoff (Gronigen), D. van Nostrand (New York, Toronto), 1952.

[60] S. C. Kleene. Recursive functionals and quantfiers of finite types. *I. Trans. Amer. Math. Soc.*, 91:1–52, 1959.

[61] U. Kohlenbach. Effective bounds from ineffective proofs in analysis: an application of functional interpretation and majorization. *The Journal of Symbolic Logic*, 57:1239–1273, 1992.

[62] U. Kohlenbach. Analyzing proofs in analysis. In *W. Hodges, M. Hyland, C. Steinhorn, J. Truss, Editors, Logic: from Foundations to Applications* at *European Logic Colloquium (Keele, 1993)*, pages 225–260. Oxford University Press, 1996.

[63] U. Kohlenbach. Mathematically strong subsystems of analysis with low rate of growth of provably recursive functionals. *Arch. Math. Logic*, 36:31–71, 1996.

[64] U. Kohlenbach. Arithmetizing proofs in analysis *in l*arrazabal, J.M., Lascar, D., Mints, G. (Editors): Logic Colloquium '96. *Springer Lecture Notes in Logic*, 12:115–158, 1998.

[65] U. Kohlenbach. Elimination of Skolem functions for monotone formulas in analysis. *Arch. Math. Logic*, 37(5-6):363–390, 1998. Logic Colloquium '95 (Haifa).

[66] U. Kohlenbach. On the arithmetical content of restricted forms of comprehension, choice and general uniform boundedness. *Ann. Pure and Applied Logic*, 95:257–285, 1998.

[67] U. Kohlenbach. On the No-Counterexample Interpretation. *The Journal of Symbolic Logic*, 64:1491–1511, 1999.

[68] U. Kohlenbach. Things that can and things that cannot be done in PRA. *Ann. Pure Applied Logic*, 102:223–245, 2000.

[69] U. Kohlenbach. Some computational aspects of metric fixed-point theory. *Nonlinear Analysis*, 61:823–837, 2005.

[70] U. Kohlenbach. Some logical metatheorems with application in functional. *Trans. Am. Math. Soc.*, 357:89–128, 2005.

[71] U. Kohlenbach. *Applied proof theory: proof interpretations and their use in mathematics*. Springer Monographs in Mathematics. Springer-Verlag, Berlin, 2008.

[72] U. Kohlenbach. On the logical analysis of proofs based on nonseparable Hilbert space theory. In *Proofs, categories and computations*, volume 13 of *Tributes*, pages 131–143. Coll. Publ., London, 2010.

[73] U. Kohlenbach. On quantitative versions of theorems due to F.E. Browder and R. Wittmann. *Advances in Mathematics*, 226(3):2764–2795, 2011.

[74] U. Kohlenbach. On the asymptotic behavior of odd operators. *Journal of Mathematical Analysis and Applications*, 382(2):615–620, 2011.

[75] U. Kohlenbach. Gödel functional interpretation and weak compactness. *Annals of Pure and Applied Logic*, 163(11):1560–1579, 2012. <ce:title>Kurt Goedel Research Prize Fellowships 2010</ce:title>.

[76] U. Kohlenbach. A uniform quantitative form of sequential weak compactness and Baillon's nonlinear ergodic theorem. *Commun. Contemp. Math.*, 14(1):1250006, 20, 2012.

[77] U. Kohlenbach and L. Leuştean. Effective metastability of Halpern iterates in cat(0) spaces. *Adv. Math.*, 231:2526–2556, 2012.

[78] U. Kohlenbach and L. Leuştean. On the computational content of convergence proofs via banach limits. *Philosophical Transactions of the Royal Society*, 370:3449–3463, 2012.

[79] U. Kohlenbach and L. Leuştean. Mann iterates of directionally nonexpansive mappings in hyperbolic spaces. *Abstr. Appl. Anal.*, (8):449–477, 2003.

[80] U. Kohlenbach and L. Leuştean. A quantitative mean ergodic theorem for uniformly convex Banach spaces. *Ergodic Theory Dynam. Systems*, 29(6):1907–1915, 2009.

[81] U. Kohlenbach and L. Leuştean. Asymptotically nonexpansive mappings in uniformly convex hyperbolic spaces. *J. Eur. Math. Soc. (JEMS)*, 12(1):71–92, 2010.

[82] U. Kohlenbach and P. Oliva. Proof Mining: A systematic way of analyzing proofs in mathematics. In *Proceedings of the Steklov Institute of Mathematics, Vol. 242*, pages 136–164, 2003.

[83] U. Kohlenbach and P. Safarik. Fluctuations, effective learnability and metastability in analysis. to appear in Ann. Pure and Applied Logic, 2014.

[84] D. Körnlein. Quantitative results for Halpern iterations of nonexpansive mappings. Preprint, 2013 submitted.

[85] G. Kreisel. On the interpretation of non-finitist proofs, part I. *The Journal of Symbolic Logic*, 16:241–267, 1951.

[86] G. Kreisel. Interpretation of analysis by means of constructive functionals of finite types. In *Constructivity in mathematics: Proceedings of the colloquium held at Amsterdam, 1957 (edited by A. Heyting)*, Studies in Logic and the Foundations of Mathematics, pages 101–128. North-Holland Publishing Co., Amsterdam, 1959.

[87] U. Krengel. *Ergodic theorems*, volume 6 of *de Gruyter Studies in Mathematics*. Walter de Gruyter & Co., Berlin, 1985. With a supplement by Antoine Brunel.

[88] A. Kreuzer. The cohesive principle and the Bolzano-Weierstraß principle. *Math. Log. Q.*, 57(3):292–298, 2011.

[89] J.-L. Krivine. Opérateurs de mise en mémoire et traduction de Gödel. *Arch. Math. Logic*, 30(4):241–267, 1990.

[90] C. Kuratowski. *Topologie Vol. I.* Warsawa, 1952.

[91] S. Kuroda. Intuitionistische Untersuchungen der formalistischen Logik. *Nagoya Math. J.*, 2:35–47, 1951.

[92] L. Leustean. A quadratic rate of asymptotic regularity for CAT(0)-spaces. *J. Math. Anal. Appl.*, 325(1):386–399, 2007.

[93] H. Luckhardt. Extensional Gödel functional interpretation. *Springer Lecture Notes in Mathematics*, vol. 306, 1973.

[94] I. Miyadera. Nonlinear mean ergodic theorems. *Taiwanese J. Math.*, 1(4):433–449, 1997.

[95] I. Moerdijk. A model for intuitionistic non-standard arithmetic. *Ann. Pure Appl. Logic*, 73(1):37–51, 1995.

[96] I. Moerdijk and E. Palmgren. Minimal models of Heyting arithmetic. *J. Symbolic Logic*, 62(4):1448–1460, 1997.

[97] E. Nelson. Internal set theory: a new approach to nonstandard analysis. *Bull. Amer. Math. Soc.*, 83(6):1165–1198, 1977.

[98] E. Nelson. The syntax of nonstandard analysis. *Ann. Pure Appl. Logic*, 38(2):123–134, 1988.

[99] P. G. Odifreddi. *Classical Recursion Theory*. Studies in Logic and the Foundations of Mathematics. Elsevier Science B.V., 1989.

[100] P. Oliva and T. Powell. A game-theoretic computational interpretation of proofs in classical analysis. *arXiv e-prints (1204.5244)*, to appear in: Gentzen Centenary volume, 2012.

[101] P. Oliva and T. Powell. On spector's bar recursion. *Mathematical Logic Quarterly*, 58(4-5):356–265, 2012.

[102] E. Palmgren. Developments in constructive nonstandard analysis. *Bull. Symbolic Logic*, 4(3):233–272, 1998.

[103] C. Parsons. Proof-theoretic analysis of restricted induction schemata (abstract). *The Journal of Symbolic Logic*, 36:361, 1971.

[104] C. Parsons. On n-quantifier induction. *The Journal of Symbolic Logic*, 37:466–482, 1972.

[105] B. D. Rouhani. Ergodic theorems for nonexpansive curves defined on general semigroups in a Hilbert space. *Nonlinear Anal.*, 44(5, Ser. A: Theory Methods):627–643, 2001.

[106] P. Safarik. On the Interpretation of the Bolzano-Weierstraß Principle Using Bar Recursion. Diplomarbeit, TU Darmstadt, 2008.

[107] P. Safarik. A quantitative nonlinear strong ergodic theorem for Hilbert spaces. *J. Math. Anal. Appl.*, 391:26–37, 2012.

[108] P. Safarik and U. Kohlenbach. On the computational content of the Bolzano-Weierstraß principle. *MLQ Math. Log. Q.*, 56(5):508–532, 2010.

[109] B. Scarpellini. A model of bar recursion of higher types. *Compositio Mathematica*, 23:123–153, 1971.

[110] K. Schütte. Beweistheorie. *Die Grundlehren der Mathematischen Wissenschaften in Einzeldarstellungen*, 103:1–355, 1960.

[111] H. Schwichtenberg. On Bar Recursion of Types 0 and 1. *The Journal of Symbolic Logic*, 44:325–329, 1979.

[112] J. R. Shoenfield. *Mathematical logic*. Association for Symbolic Logic, Urbana, IL, 2001. Reprint of the 1973 second printing.

[113] W. Sieg. Fragments of arithmetic. *Ann. Pure Appl. Logic*, 28:33–71, 1985.

[114] S. G. Simpson, editor. *Subsystems of Second Order Arithmetic*. Perspectives in Mathematical Logic. Springer-Varlag, Berlin · Heidelberg · New York, 1999.

[115] S. G. Simpson. *Subsystems of second order arithmetic*. Perspectives in Mathematical Logic. Springer-Verlag, Berlin, 1999.

[116] S. G. Simpson and K. Yokoyama. A nonstandard counterpart of wwkl. *Notre Dame Journal of Formal Logic*, 52(3):229–243, 2010.

[117] C. Spector. Provably recursive functionals of analysis: a consistency proof of an analysis by extension of principles formulated in current intuitionistic mathematics. In *Recursive function Theory, Proceedings of Symposia in Pure Mathematics, vol. 5 (J.C.E. Dekker (ed.)), AMS, Providence, R.I.*, pages 1–27, 1962.

[118] T. Streicher and U. Kohlenbach. Shoenfield is Gödel after Krivine. *Mathematical Logic Quarterly*, 53(2):176–179, 2007.

[119] T. Streicher and B. Reus. Classical logic, continuation semantics and abstract machines. *Journal of Functional Programming*, 8(06):543–572, 1998.

[120] W. W. Tait. Constructive reasoning. In *Logic, methodology and the philosophy of science. III, (B. van Rootselaar and J. F. Staal (ed.)), North-Holland, Amsterdam*, pages 185–199, 1976.

[121] K. Tanaka. Non-standard analysis in wkl0. *Mathematical Logic Quarterly*, 43(3):396–400, 1997.

[122] T. Tao. Norm convergence of multiple ergodic averages for commuting transformations. *Ergodic Theory and Dynamical Systems*, 28:657–688, Apr. 2008.

[123] T. Tao. *Soft analysis, hard analysis, and the finite convergence principle*, page 298pp. American Mathematical Soc., 2008.

[124] M. Toftdal. Calibration of ineffective theorems of analysis in a constructive context. Master's thesis, Aarhus Universitet, 2004.

[125] A. S. Troelstra, editor. *Metamathematical investigation of intuitionistic arithmetic and analysis.* Lecture Notes in Mathematics, Vol. 344. Springer-Verlag, Berlin, 1973.

[126] A. S. Troelstra. Note on the Fan Theorem. *The Journal of Symbolic Logic*, 39:584–596, 1974.

[127] A. S. Troelstra and D. van Dalen. *Constructivism in mathematics. Vol. II*, volume 123 of *Studies in Logic and the Foundations of Mathematics*. North-Holland Publishing Co., Amsterdam, 1988. An introduction.

[128] J. von Neumann. Proof of the quasi-ergodic hypothesis. In *Proc. Nat. Acad. Sci. USA*, 1932.

[129] R. Wittmann. Mean ergodic theorems for nonlinear operators. *Proc. Amer. Math. Soc.*, 108:781–788, 1990.

[130] R. Wittmann. Approximation of fixed points of nonexpansive mappings. *Arch. Math.*, 58:486–491, 1992.

[131] H.-K. Xu. Iterative algorithms for nonlinear operators. *J. London Math. Soc.*, 66:240–256, 2002.

[132] M. Ziegler. Real hypercomputation and continuity. *Theory of Computing Systems*, 41:177–206, 2007.